PRACTICAL APPLICATIONS OF RADIOACTIVITY AND NUCLEAR RADIATIONS

This book introduces selected examples from the numerous applications of radioisotopes and ionising radiations to engineering and environmental sciences and technologies. In addition, it serves teachers and students as an introductory course in nuclear sciences.

The early chapters introduce the properties of radionuclides, radioactive sources, ionising radiations, detectors and instrumentation, and how they are used. Also described are the methods for obtaining accurate countrate measurements allowing for the statistics of radioactive decays. Later chapters introduce applications to mining, mineral processing, petroleum refining, contaminant transport, borehole logging, fluid flow measurements and to the assessment of sites for radioactive waste disposal. The fact that different radiotracers can be used to separately label and study different components of complex systems is an outstanding example of their versatility.

This book will be of particular interest to scientists, technologists, teachers and students, helping them to work with radioisotopes safely, efficiently and reliably.

Since the 1960s, Drs Lowenthal and Airey served on the research staff of the Australian Nuclear Science and Technology Organisation, ANSTO (formerly the Australian Atomic Energy Commission, AAEC), where Dr Airey continues to hold a senior position.

DR LOWENTHAL was the leader of a small group which established the Australian Standards of Radioactivity. In the mid-1970s he became the Australian Representative on the International Committee for Radionuclide Metrology (an independent organisation of specialists) and was invited to represent Australia on the Consultative Committee for Standards for the Measurement of Ionising Radiations on the International Committee for Weights and Measures, Paris. He has published some 50 research papers with other members of the group and for six years was honorary Australasian Editor of the *International Journal of Applied Radiation and Isotopes.* He is currently Honorary Consultant for Nuclear Medicine at the Royal Prince Alfred Hospital and an Honorary Consultant in the School of Mechanical and Manufacturing Engineering at the University of New South Wales. It was largely for his contributions to radionuclide metrology that he was named (in 1994 in Paris), Chevalier de l'Ordre National du Merite and, in 1999, was made a Member of the General Division of the Order of Australia (AM).

DR AIREY is a physical chemist who has been extensively involved in research into the applications of radioactivity and radiation to industry and the environment. From 1986 to 1990, he was posted to the International Atomic Energy Agency where he coordinated a range of programs involving these applications in Asia and the Pacific. He is currently Australian National Counterpart for selected IAEA projects. Prior to his appointment to the IAEA and on his return, Dr Airey was responsible for coordinating an international OECD Nuclear Energy Agency project concerned with the evaluation of uranium deposits as natural analogues of radioactive waste repositories. From 1992 to 1999, he was ANSTO's representative on the independent Safety Review Committee. Dr Airey is a past president of the Australian Nuclear Association and has published over sixty research papers and technical reports.

PRACTICAL APPLICATIONS OF RADIOACTIVITY AND NUCLEAR RADIATIONS

An introductory text for engineers, scientists, teachers and students

G. C. LOWENTHAL

University of New South Wales

P. L. AIREY

Australian Nuclear Science and Technology Organisation

CAMBRIDGE
UNIVERSITY PRESS

CAMBRIDGE UNIVERSITY PRESS
Cambridge, New York, Melbourne, Madrid, Cape Town, Singapore, São Paulo

Cambridge University Press
The Edinburgh Building, Cambridge CB2 2RU, UK

Published in the United States of America by Cambridge University Press, New York

www.cambridge.org
Information on this title: www.cambridge.org/9780521553056

First published 2001
This digitally printed first paperback version 2005

A catalogue record for this publication is available from the British Library

ISBN-13 978-0-521-55305-6 hardback
ISBN-10 0-521-55305-9 hardback

ISBN-13 978-0-521-01980-4 paperback
ISBN-10 0-521-01980-X paperback

Contents

Illustrations

Tables

Foreword

I welcome the publication of the book *Practical Applications of Radioactivity and Nuclear Radiations – An introductory text for engineers, scientists, teachers and students* by Drs Gerhart Lowenthal and Peter Airey.

This book aims to provide readily accessible information on the applications of nuclear science and technology to industry, the environment and scientific research. It is tailored to students and non-specialists who seek to be informed, and may eventually wish to contribute to this world-wide endeavour. Emphasis is placed on the science underpinning the applications.

An objective of international bodies such as the Forum for Nuclear Cooperation in Asia (FNCA) and the International Atomic Energy Agency (IAEA) is to extend the benefits of nuclear related technologies across national boundaries. One of the authors (P.A.) was involved for a number of years in this endeavour through the IAEA.

I am pleased to endorse this book as a further step in this on-going quest.

Dr Sueo Machi
FNCA Coordinator, Japan
Former Deputy Director General, Nuclear Science and Applications, IAEA

The discovery of X rays and radioactivity in the 1890s had a profound effect on the century which was to follow. After more than one hundred years of endeavour which has been both marred by conflict and enriched by high achievement, we enter the new millennium with a more mature understanding of the benefits that nuclear technology can bring to mankind. These extend to medicine, agriculture, industry, the environment and the exploration for natural resources.

The potential for nuclear and isotope techniques to contribute to the advance of scientific knowledge appears unbounded. As shown by examples in the book, these techniques contribute to our understanding of the structure of modern materials, the dynamics of biological systems, the complexity of many ecosystems and the dating of terrestrial processes from the recent past back almost to the formation of the earth.

This book is written for students and non-specialist scientific workers and aims to raise the awareness of the practical application of isotopes and radiation to industry and the environment. I support its publication. I believe that people everywhere will continue to benefit from the safe applications of nuclear science and technology in the years ahead.

Professor Helen M. Garnett
Chief Executive
Australian Nuclear Science and Technology Organisation.

Preface

Radionuclides and the emitted radiations have long been applied routinely throughout all branches of engineering and the technologies to obtain useful results, many of which could not have been obtained by other means.

As with the application of any other tool or technique, problems can be encountered. Practitioners working with radioactivity face health risks but long-standing records show that, overall, risks in the nuclear industries have been consistently smaller than those faced by workers in most other industries. This is so not least thanks to easily followed, legally backed precautions, developed over the decades to ensure safe operating conditions during all nuclear radiation applications carried out within the common sense rules devised for that purpose.

This book was written for workers and students as yet largely unfamiliar with the nuclear sciences and with the advantages in numerous fields which quickly become apparent on employing nuclear radiations. The potential of nuclear science and engineering for enhancing, e.g. the effectiveness of nuclear power production and of radioactive tracers is far from exhausted. Today's beginners could have a highly rewarding way ahead of them.

A useful overview of the contents can be readily obtained by scanning the chapter and section headings in the table of Contents. The bibliography contains over 120 references to assist practitioners looking for more detailed and/or more specialised information than could be included here. The latest information about nuclear data and specialist techniques is available via the Internet, with web sites and other comments listed in Appendix 3.

One of us (P.A.) spent four years with the International Atomic Energy Agency (Vienna, Austria) involved with the applications of radioisotopes and radiation to industry, medicine, agriculture and the environment in Asia and the Pacific. The knowledge and dedication of many colleagues both at the

IAEA and in its Member States experienced during these years and since has greatly benefited the writing of this book.

We gratefully acknowledge the contributions from many colleagues who read sections of this text, making constructive suggestions and pointing out inadequate explanations and errors. There is Dr D.D. Hoppes, formerly of the USA National Institute of Standards and Technology, Gaithersburg (NIST), USA and Dr J.S. Charlton, former General Manager, Tracerco (Australasia). To our regret we can no longer thank Dr A. Rytz, formerly of the International Bureau of Weights and Measures (BIPM), Paris, France, who died early in 1999 and is sadly missed. Overseas colleagues who helped us materially with abstracts from their publications were Dr W.B. Mann (NIST, USA), Dr K. Debertin (Germany), and Drs T. Genka, H. Miyahara, Y. Kawada, Y. Hino and Professor T. Watanabe (Japan).

We are deeply indebted to colleagues at the Australian Nuclear Science and Technology Organisation (ANSTO) of Lucas Heights, NSW, who have been engaged on applications of nuclear radiations for two and more decades and assisted us in numerous ways. In particular, advice on X and gamma ray spectroscopy came from Dr D. Alexiev and Mr A.A. Williams. Special thanks are due to the radiotracer team at ANSTO, including Mr T. Kluss, Dr C. Hughes and Mr G. Spelman. Assistance came also from Dr B. Perczuk, School of Physics, University of New South Wales.

Other helpful suggestions came from Mr A. Fleischman and Mrs J. Towson of the Australian Radiation Protection Society and Mr J.S. Watt, Chief Research Scientist in the Commonwealth Scientific and Industrial Research Organisation. We also acknowledge valuable technical assistance from Mr S. Eberl of the Royal Prince Alfred Hospital (RPAH), Sydney, Mr A.W.L. Hu, Miss Adrienne Walker and Mr W. Hu all of the School of Mechanical and Manufacturing Engineering, University of New South Wales.

A very special "thank you" goes to the extremely helpful and valuable secretarial assistance from Mrs Diane Augee, also at the School of Mechanical and Manufacturing Engineering, UNSW. Mrs Augee could not protect us from our oversights though she did so in a good many cases. She always made sure that the text was throughout well arranged, clearly expressed and correctly spelled.

We acknowledge with most sincere gratitude support over the many years while writing this book from the Chief Executive, ANSTO (Professor H.M. Garnett) and the Director, Environmental Division, ANSTO (Professor A. Henderson-Sellers), and from the School of Mechanical and Manufacturing Engineering, UNSW (Professor B.E. Milton and later Professor K.P. Byrne) and the Dean of Engineering (Professor M.S. Wainwright). G.L. also received substantial support at the Department of Positron Emission Tomography and Nuclear Medicine, Royal Prince Alfred Hospital, Sydney (Professor M.J. Fulham and earlier Professor J. Morris). Finally, our apologies go to colleagues whom we failed to mention, to our readers who will have to put up with remaining errors, and to our wives who put up with the demands the book made on our time and assisted us in every way.

Chapter 1

Atoms, nuclides and radionuclides

1.1 Introduction

1.1.1 Radioactivity, from the 1890s to the 1990s

Radioactivity is a characteristic of the nuclei of atoms. The nuclei, and with them the atoms as a whole, undergo spontaneous changes known as radioactive or nuclear transformations and also as decays or disintegrations. The energy released per nuclear transformation and carried away as nuclear radiation is, as a rule, some 10^3 to 10^6 times greater than the energy released per atom involved in chemical reactions.

Radioactivity was discovered in 1896 by the Frenchman H. Becquerel. The discovery occurred while he was experimenting with phosphorescence in compounds of uranium, an investigation aiming only at knowledge for its own sake. However, practical applications of radioactivity appeared not long after its discovery and have multiplied ever since.

Radioactivity could not have been discovered much before 1896 because at naturally occurring intensities it is undetectable by the unaided human senses. The photographic technique which contributed to its discovery was not adequately developed until well into the nineteenth century. But by the end of that century it played a major part in two discoveries which changed the path of science and of history: Röntgen's discovery of X rays in Germany in late 1895, followed by Becquerel's discovery of radioactivity in France in early 1896. These completely unexpected events opened the doors to totally new physical realities, to the emerging world of the nuclei of atoms and the high-energy nuclear radiations emitted by these nuclei.

Following Becquerel's discovery, it took about 35 years of intense scientific work before radioactive atoms could be produced from stable atoms by man-made procedures. Methods have now been developed to produce over 2000

1

Figure 1.1. Periodic table of the chemical elements

IA	IIA	IIIA 3	IVA 4	VA 5	VIA 6	VIIA 7	VIIA	VIIA	VIIA	IB	IIB	IIIB	IVB	VB	VIB	VIIB	VIIIB
1 **H** 1.008																	2 **He** 4.00
3 **Li** 6.94	4 **Be** 9.01											5 **B** 10.81	6 **C** 12.01	7 **N** 14.01	8 **O** 16.00	9 **F** 19.00	10 **Ne** 20.18
11 **Na** 22.99	12 **Mg** 24.31											13 **Al** 26.98	14 **Si** 28.09	15 **P** 30.97	16 **S** 32.06	17 **Cl** 35.45	18 **Ar** 39.95
19 **K** 39.10	20 **Ca** 40.08	21 **Sc** 44.96	22 **Ti** 47.90	23 **V** 50.94	24 **Cr** 52.00	25 **Mn** 54.94	26 **Fe** 55.85	27 **Co** 58.93	28 **Ni** 58.71	29 **Cu** 63.55	30 **Zn** 65.38	31 **Ga** 69.72	32 **Ge** 72.59	33 **As** 74.92	34 **Se** 78.96	35 **Br** 79.90	36 **Kr** 83.80
37 **Rb** 85.47	38 **Sr** 87.62	39 **Y** 88.91	40 **Zr** 91.22	41 **Nb** 92.91	42 **Mo** 95.94	43 **Tc** 98.91	44 **Ru** 101.07	45 **Rh** 102.91	46 **Pd** 106.4	47 **Ag** 107.87	48 **Cd** 112.40	49 **In** 114.82	50 **Sn** 118.69	51 **Sb** 121.75	52 **Te** 127.60	53 **I** 126.90	54 **Xe** 131.30
55 **Cs** 132.91	56 **Ba** 137.34	57 **La** 138.91	72 **Hf** 178.49	73 **Ta** 180.95	74 **W** 183.85	75 **Re** 186.2	76 **Os** 190.2	77 **Ir** 192.22	78 **Pt** 195.09	79 **Au** 196.97	80 **Hg** 200.59	81 **Tl** 204.37	82 **Pb** 207.2	83 **Bi** 208.98	84 **Po** [209]	85 **At** [210]	86 **Rn** [222]
87 **Fr** [223]	88 **Ra** [226]	89 **Ac** [227]	104 **Rf** [261]	105 **Db** [262]	106 **Sg** [263]	107 **Bh** [264]	108 **Hs** [265]	109 **Mt** [268]									

58 **Ce** 140.12	59 **Pr** 140.91	60 **Nd** 144.24	61 **Pm** [145]	62 **Sm** 150.4	63 **Eu** 151.96	64 **Gd** 157.25	65 **Tb** 158.93	66 **Dy** 162.50	67 **Ho** 164.93	68 **Er** 167.26	69 **Tm** 168.93	70 **Yb** 173.04	71 **Lu** 174.97
90 **Th** 232.04	91 **Pa** 231.04	92 **U** 238.03	93 **Np** 237.05	94 **Pu** [244]	95 **Am** [243]	96 **Cm** [247]	97 **Bk** [247]	98 **Cf** [251]	99 **Es** [254]	100 **Fm** [257]	101 **Md** [258]	102 **No** [255]	103 **Lr** [256]

Figure 1.1.   The periodic table of the chemical elements as established by the early 1990s listing all elements identified and named by that date. The entries list the Z number, the symbol and the atomic weight. The lanthanum and actinide series are parts of periods VI and VII. For further comments see Sections 1.2.1, 1.2.2, 1.5.3.

radioisotopes of the chemical elements listed in the periodic table (Figure 1.1). However, the number of radionuclides with well known characteristics is much lower than that.

Charts of Nuclides published by nuclear research centres (see Section 4.2.1) list some 280 stable isotopes and commonly over 1800 radioisotopes, depending on the criteria used for the selection. Also, the number is increasing with time. There are on average four to eight radioisotopes for elements up to carbon (element 12 in the periodic table), 25 to 35 radio-isotopes for each of the heavier of the stable elements, say between tantalum and bismuth (numbers 73 to 83), and there are the isotopes of the naturally occurring and man-made radioelements from polonium upwards. At present between 100 and 200 radionuclides are used on a regular basis, supporting a world-wide industry which continues to grow.

1.1.2   On the scope and content of this text

Clearly, large numbers of radionuclides are available and readers will be advised where to look for their characteristics. Routine applications require knowledge of the relevant physical and chemical properties of rarely more than a few of these radioisotopes, together with basic knowledge about radioactivity as offered in this or similar textbooks. Given these foundations, researchers, technologists, teachers and students will find nuclear radiation applications helpful assets for solving many of their day-to-day problems.

In working through this book experimenters seeking more specialised information will be referred to publications that deal with specific issues in more detail than can be covered in this book. It may be helpful to know in advance that many of these references cite the following publications.

Alfassi, Ed. (1990), *Activation Analysis*
Charlton, Ed. (1986), *Radioisotope Techniques for Problem Solving in Industry*
Debertin and Helmer (1988), *γ and X ray Spectrometry with Semiconducting Detectors*
Heath (1974), *Gamma Ray Spectrum Catalogue*
IAEA (1990), *Guidebook on Radioisotope Tracers in Industry*
L'Annunziata, Ed. (1998), *Handbook of Radioactivity Analysis*
Longworth, Ed. (1998), *The Radiochemical Manual*
Mann *et al.* (1991), *Radioactivity Measurements, Principles and Practice*
NCRP (1985), *A Handbook of Radioactivity Measurement Procedures*
Peng (1977), *Sample Preparation in Liquid Scintillation Counting*
Seltzer and Berger (1982), *Energy Losses and Ranges of Electrons and Positrons.*

Full titles and publishers are listed in the References at the end of this book.

As can be seen in the Contents, Chapters 1 to 6 deal principally with introductory material to the science of radioactivity while Chapters 7 to 9 introduce a large range of applications, briefly describing how they can be used. The two parts are interdependent. Descriptions of applications contain references to the earlier sections of the book where it is pointed out how to use radioactivity, while descriptions of the rules of radioactivity contain references to sections in Chapters 7 to 9 where these rules are put to use.

To conclude this introductory section, data illustrating the extent of radionuclide applications on a national and international level are now presented.

1.1.3 Joining a large scale enterprise

Nuclear power and nuclear radiation applications

Nuclear power for commercial electricity production first became available during the mid-1950s. During the mid-1990s, a total of 470 nuclear power reactors produced 17% of the world's commercially distributed electricity (Hore-Lacy, 1999, Table 6). Nuclear power production could expand as rapidly as it did because of the wide availability of high levels of expertise and training in nuclear science and engineering which led to a similarly high rate of expansion of nuclear radiation applications. It was no accident that nuclear power generation and other nuclear radiation applications advanced together.

Figures from Japan

The figures below were published by members of the Japan Radioisotope Association (Umezawa *et al.*, 1996). Quoting from that report, it appears that in 1993 the number of facilities in Japan using radionuclides to a significant extent reached 5509, comprising:

Medical facilities	1385
Educational organisations	411
Research establishments	896
Industrial undertakings	1844
Other organisations	973

Radionuclide applications in industrial undertakings include large radiation processing facilities for food irradiation, the sterilisation of medical supplies and other commercial irradiation plant. There are also the figures referring to the types and quantities of radionuclides employed for non-medical and medical purposes, again for 1993. The total applies to radionuclides from 80 different elements.

First the number of orders, to the nearest thousand, for the purchase of non-medical applications:

Labelled compounds	63 000
Processed radionuclides	7 000

Coming to medical applications, the number of injections of radionuclides made during 1993 for diagnostic purposes exceeded 70 million. Since then most of the stated applications increased in number to a significant extent.

Among the major users of radioisotope applications it appears to be only in Japan where figures are published in such a way that they can be easily and concisely quoted. Other major users of nuclear techniques, notably the USA, produce, import and use radionuclides in a large number of independent installations. To quote these figures, and notably the role played by imported radionuclides, with sufficient accuracy requires more space than can be spared in this book.

The role of research reactors

Radionuclides for industrial and similar applications are not produced in nuclear power reactors, but in so-called research reactors. These were designed as essential tools for research in the nuclear sciences and for the production of radionuclides. The operation of research reactors will be briefly described in Sections 1.4.1 and 1.4.2.

Currently there are 280 research reactors operating in 54 countries (Hore-Lacy, 1999, p. 30), figures which demonstrate the large volume of applications and the fact that facilities for these applications are in constant operation in practically all industrialised countries around the globe.

1.2 An historic interlude: from atoms to nuclei

1.2.1 When atoms ceased to be atoms

During the fifth century BC Greek natural philosophers used the laws of logic to convince themselves and many later philosophers that matter must consist of infinitely small, hard spheres which they called atoms, the Greek word for uncuttable. This hypothesis remained without experimental backing for some 2300 years. It was only early in the nineteenth century that the English chemist John Dalton could revive the Greek model of the atom using experimental evidence to theorise that the atoms of different chemical elements differed in mass by integral multiples. Also, when atoms of different elements combine they do so in ratios of whole numbers and not otherwise.

By the 1860s Dalton's pioneering work had resulted in the discovery of over 60 chemical elements and sufficient new knowledge to lead to the formulation of what became known in due course as the periodic table of the elements proposed by the Russian chemist D.I. Mendeleyev and, independently by a German chemist L. Mayer. Figure 1.1 shows the present state of the table.

To add just a few words on its history, Mendeleyev published his arrangement of chemical elements in 1869. He had ordered them by their atomic weights, which were reasonably well known. Also, it was his idea to take care of the many similarities and differences in the properties of the elements by subdividing them into periods and groups. Beginning with about 60 elements in 1869, subsequent discoveries of elements were greatly aided by Mendeleyev's analysis. However, it was not until the coming of quantum theory during the late 1920s that Mendeleyev's periods and groups could be explained theoretically. A few explanatory comments on the periodic table will follow in Sections 1.2.2 and 1.5.3.

Although the theory behind Mendeleyev and Mayer's periodic table proved remarkably perceptive (Hey and Walters, 1987, Ch. 6), no-one anticipated the radically new knowledge which was to come as the nineteenth century ended.

The first totally inexplicable event occurred in late 1895 when W.C. Röntgen, Professor of Physics at the University of Würzburg, Germany, discovered the aptly named X rays, i.e. rays of unknown origin. The second discovery was equally inexplicable. It was made early in 1896 at the University of Paris, France, when Professor H. Becquerel discovered that the well known element uranium emits mysteriously penetrating radiations similar to X rays, a property later named radioactivity.

With the new century came two, truly revolutionary theories: quantum theory, the first stage of which was published in 1900 by Professor Planck in Berlin, Germany, and the special relativity theory published by Albert Einstein in 1905 and predicting, among others, that energy has mass (Section 1.3.2). These discoveries gave a radically new turn to the physical sciences of the twentieth century.

Following on from Becquerel's work, his student Marie Curie discovered (in 1898) two other radioactive elements she named polonium (after Poland, her country of birth) and radium (the radiator). She also identified the α and β radiations (so named by her) as charged particles emitted by radium and its daughters. A year earlier, in 1897, J.J. Thompson, a professor at Cambridge University, England, had discovered the electron which he called the atom of electricity and which was soon recognised as basic to all electric phenomena.

The most important advances involving radioactivity were made by the New Zealander Ernest Rutherford, then Professor of Physics at Manchester University, England.

Starting their researches during the first decade of the twentieth century, Professor Rutherford and his student-collaborators H. Geiger and E. Marsden used the α particles emitted by radium as projectiles to demonstrate that, although nearly all of them readily penetrated a very thin metal foil, a small fraction was actually backscattered by up to 180 degrees (Hey and Walters, 1987, Ch. 4).

These results were incompatible with the accepted theory of atomic structure due mainly to Professor Thompson, according to which the back-scatter of α particles from thin metal foils was ruled out. Rutherford then theorised that the observed very low backscatter rate could be explained if atoms consist of extremely small and dense positively charged particles that he called protons, surrounded, at a relatively very large distance, by an equal number of very light negatively charged electrons (so ensuring electrical neutrality), but otherwise empty of matter. To explain the observed back-scatter rate, the ratio of the diameter of the atom to that of its nucleus had to be of order 10^4. Since atomic diameters had been estimated as of order 10^{-10} m, the diameters of the nuclei of atoms would be as small as 10^{-14} m. These results were published in 1911, but they were totally inconsistent with existing knowledge and few researchers were willing to accept them.

Notwithstanding his successes, Rutherford was unable to explain how a nucleus consisting only of positively charged particles could avoid being torn apart by Coulomb repulsion which had to be extremely strong in so tiny a volume. He also could not explain why electrons could remain in stable orbits instead of spiralling towards the oppositely charged protons as demanded by Maxwell's electromagnetic theory, the validity of which had been tested in innumerable experiments.

Rutherford, Bohr, Heisenberg, Chadwick, Dirac and other well known physicists, most of whom were awarded Nobel Prizes, worked hard throughout the 1920s and 1930s until the properties of atoms and of their decay could be explained by the new quantum and relativity theories (Hughes, 2000).

1.2.2   The atomic nucleus

The just stated discoveries opened the way to the present knowledge that atomic nuclei consist of protons and neutrons known collectively as nucleons. It is also recognised that each nucleon consists of three smaller particles known as quarks, but the quark structure of nuclei will be ignored in what follows.

The elementary building blocks of nuclei are the positively charged protons (p), the charge being a single elementary unit, 1.602×10^{-19} coulomb (C), and the electrically uncharged neutrons (n). The number of protons in the nucleus determines its chemical nature, i.e. the chemical element to which the atoms belong. It is known as the atomic number of the element, designated Z, and serves as the order number in the periodic table (Figure 1.1).

Neutrons and protons are almost equal in mass (m), the mass ratio m_n/m_p being 1.0014, and they account for 99.95% of the inertial mass of the atom. The atomic electrons account for barely 0.05% of the inertial mass but they move within a volume some 10^{12} times larger than the volume of the nucleus, not along defined orbits as do planets but subject to probability-based relations derivable from quantum mechanics. The volume of the atom is practically empty of inertial matter, but it is permeated by intense electric fields between the charged particles. The force inside the nucleus which holds its components together is of an extremely short range ($\approx 10^{-13}$ m) and is known as the strong nuclear force.

Chemical interactions between elements are governed by forces involving the outer electrons of atoms, i.e. those least strongly bound to their nuclei. In contrast, the laws of radioactivity are determined by intra-nuclear forces which are many orders of magnitude higher than those involved in chemical interactions and well shielded from these interactions. This discovery helped to explain the almost complete independence of radioactive decay rates from all other properties of the radioactive material such as temperature, pressure or state of aggregation.

1.3 Nuclei, nuclear stability and nuclear radiations

1.3.1 The birth of isotopes

Dalton's ideas of integral atomic weights also ran into difficulties. As weighing techniques improved it was discovered that some 20% of the elements have atomic weights that are very nearly integers but not quite. For example, the atomic weight of phosphorus is not 31.00 but 30.97, close to 0.1% smaller than expected. The large majority of atomic weights differ from the nearest integer by amounts which are nearer to half a mass unit than to zero. It was later realised that the large majority of elements did not consist of a single species of atoms but of two and more species, each representing a well defined proportion of the total.

For elements consisting of more than one species of atom, each atom has the same number of protons in its nucleus. However, and this was only clarified with the discovery of the neutron in 1932, these species of atoms differ in the number of neutrons in their nuclei. This difference causes a difference in their atomic mass but has virtually no effect on their chemistry. Species of atoms that belong to the same element but differ in nuclear mass and so in atomic weight were labelled isotopes (derived from Greek and meaning same position) of the element in question.

Only about 20 elements consist of a single stable isotope as does phosphorus. The atoms of the other elements consist of several isotopes, with tin ($Z = 50$) consisting of ten isotopes. The atomic weights of these atoms are the averages of the weights of the isotopes weighted in accordance with their relative abundances. As could be expected, none of these averages is an integer (Figure 1.1).

1.3.2 Mass–energy conversions and the half life

Another characteristic of the nucleus is its mass number, designated A. It is the sum of the number of protons and neutrons in the nucleus with each of them assigned unit mass. The mass of the electrons is disregarded as is the small difference in mass between protons and neutrons. The mass number is then necessarily an integer given by $A = p + n = Z + n$ since $p = Z$ (Section 1.2.2). It is written as the raised prefix to the chemical symbol of the element, ^{A}Z.

Returning to phosphorus, its atoms each contain 15 protons and 16 neutrons, i.e. $A = 15 + 16 = 31$, written ^{31}P. But its atomic weight is 30.97 as noted earlier, a difference which is of fundamental importance.

Measurements of atomic weights are now made relative to the mass of the carbon isotope ^{12}C which is set as equal to 12 atomic mass units (amu). The energy equivalent of an atomic mass unit is calculated using the Einstein equation E (joules, J) $= mc^2$ where m is expressed in kilograms and c is closely equal to 3×10^8 m/s, the speed of light in vacuum. It can then be shown that 1 amu = 932 million electronvolts (MeV) as its energy equivalent. The electron-volt (eV) is related to the joule by 1 eV $= 1.602 \times 10^{-19}$ J.

If the 0.03 mass units which make up the difference between the mass number $A = 31$ and the atomic weight of phosphorus are expressed in energy terms (~30 MeV), one obtains the energy that holds the 31 nucleons in the phosphorus nucleus together. This energy is known as the nuclear binding energy (E_{Bi}) for phosphorus and is obtained at the expense of a small fraction of the inertial mass of the nucleons.

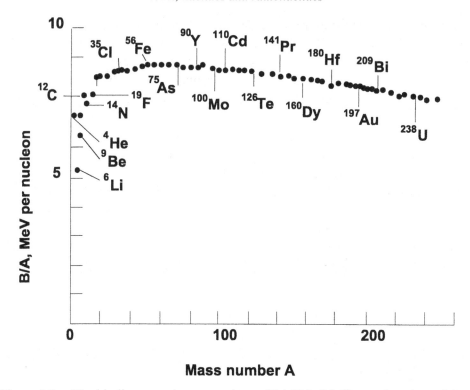

Figure 1.2.   The binding energies per nucleon, B/A (B in MeV) as a function of the mass number $A(=p+n)$. Note the strongly bound helium nucleus (α particle). Adapted from Hey and Walters (1987, Ch. 5).

The magnitude of E_{Bi} increases with the mass number. However, the binding energy per nucleon, E_{Bi}/A, increases to a maximum near $Z = 26$ (iron, Figure 1.1) where $A \approx 50$. It then decreases, being some 15% smaller near $Z = 92$ (uranium) where $A \approx 238$, than near $A \approx 50$ (Figure 1.2).

The release of nuclear binding energy provides the energy for all radioactive transformations (Section 1.5.4). Also, the fact that E_{Bi}/A goes through a maximum near $A = 50$ permits the release of very large amounts of energy in two situations: during the fusion of light nuclei, notably of hydrogen nuclei, and during the fissioning of a heavy nucleus, notably that of the uranium isotope ^{235}U. Uranium ($Z = 92$) has numerous other isotopes, none of which is stable (Section 1.3.4) and two being of interest here: ^{235}U and ^{238}U with respectively $235 - 92 = 133$ neutrons and $238 - 92 = 136$ neutrons in their nuclei (Section 1.4.1).

Another nucleus to be noted (Figure 1.2) is that of helium (2p + 2n), better known as an α particle. The figure shows that helium is much more strongly bound than its neighbours, so helping to explain why its protons and neutrons are not emitted singly from radioactive nuclei but combined into a single particle, the α particle.

Another measure of the instability of a radionuclide is its half life, the magnitude of which is a characteristic of each radionuclide with no two having the same value. It is defined as follows. On starting at time t_1 with a sample of N radioactive atoms of the same radionuclide, let this number be reduced by radioactive decay to $N/2$ at time t_2. The half life, written $T_{1/2}$, is defined as equal to the interval $t_2 - t_1$. Measured half lives of radionuclides range from small fractions of a microsecond to 10^{20} years and even longer.

1.3.3   From natural to man-made radioisotopes

In the first 35 years following the discovery of radioactivity the only radio-nuclides available for experimentation were those occurring naturally. This changed with the coming of high-voltage accelerators during the early 1930s. It then began to be realised that all elements could have radioactive isotopes, which came to be known as radioisotopes.

The first experimentally produced radioisotopes were prepared during the early 1930s by Frederic and Irene Joliot-Curie (the daughter of Marie Curie) at the University of Paris. They caused α particle projectiles to interact with the atoms in a thin foil of boron so producing atoms of nitrogen-13 with a half life of 10 minutes, a discovery for which the 1934 Nobel Prize in Chemistry was awarded, the third for the Curie family.

To penetrate into positively charged nitrogen nuclei, the positively charged α particles have to overcome strong Coulomb repulsion, making this process very inefficient. Meanwhile, in February 1932 in Rutherford's laboratory, J. Chadwick had discovered the neutron. It was not long thereafter that Enrico Fermi, Professor of Physics at the University of Rome, Italy, realised that neutrons, being uncharged, would be unaffected by the Coulomb barrier and so could penetrate into atomic nuclei to effect radioactivation with much higher efficiencies than was possible for α particles.

Experimenting throughout the mid-1930s, Fermi and his collaborators demonstrated that neutrons emitted from a radium–beryllium source (Section 1.4.4), could produce radioisotopes of most stable elements (see Eq. (1.1), Section 1.4.3). To their surprise, radioactivation was more efficient when caused by low-energy (slow) than by high-energy (fast) neutrons. However, they just failed to discover that neutron penetration into uranium did not cause uranium atoms to turn into atoms of a neighbouring element, but caused some of its nuclei to break up, fission, with the release of large amounts of energy.

Z	Element	12	13	14	15	16	17	18	19	20
21	Sc 44.96 σ 26.5								Sc 40 0.18 s β⁺ γ	Sc 41 0.60 s β⁺ γ
20	Ca 40.08 σ 0.43					Ca 36 0.1 s β⁺ γ	Ca 37 0.18 s β⁺ γ	Ca 38 0.44 s β⁺ γ	Ca 39 0.86 s β⁺	Ca 40 96.94 σ 0.40
19	K 39.10 σ 2.10					K 35 0.19 s β⁺ γ	K 36 0.34 s β⁺ γ	K 37 1.22 s β⁺ γ	K 38 7.6 m β⁺ γ	K 39 96.26 σ 2.0
18	Ar 39.95 σ 0.68			Ar 32 98 ms β⁺ γ	Ar 33 0.17 s β⁺ γ	Ar 34 0.84 s β⁺ γ	Ar 35 1.78 s β⁺ γ	Ar 36 0.34 σ 5.6	Ar 37 35.0 EC	Ar 38 0.06 σ 0.8
17	Cl 35.45 σ 33.2			Cl 31 150 ms β⁺	Cl 32 0.29 s β⁺ γ	Cl 33 2.5 s β⁺ γ	Cl 34 32.0 m β⁺ γ	Cl 35 75.8 σ 43	Cl 36 3×10⁵y β⁻ EC	Cl 37 24.2 σ 0.43
16	S 32.06 σ 0.54	S 28 125 ms β⁺	S 29 0.19 s β⁺ γ	S 30 1.18 s β⁺ γ	S 31 2.58 s β⁺ γ	S 32 95.0 σ 0.53	S 33 0.75 σ 0.46	S 34 4.21 σ 0.29	S 35 87.5 d β⁻	S 36 0.02 σ 0.23
15	P 30.97 σ 0.18	P 27 0.26 s β⁺	P 28 0.27 s β⁺ γ	P 29 4.1 s β⁺ γ	P 30 2.50 m β⁺ γ	P 31 100 σ 0.16	P 32 14.26 d β⁻	P 33 25.3 d β⁻	P 34 12.4 s β⁻ γ	P 35 47.4 s β⁻ γ
		12		14		16		18		20

→ n

Figure 1.3. A small, simplified section from a *Chart of Nuclides* (CoN, 1988). The stable nuclides (black squares) are shown with their isotopic fractions and thermal neutron cross sections (σ barn). Radioisotopes (white squares) are shown with their half lives and decay modes.

Fission was discovered late in 1938 by O. Hahn and F. Strassmann in Berlin, Germany. This discovery led, in 1942, to the construction of the first nuclear fission reactor (in Chicago, USA) under the direction of Professor Fermi who had been awarded a Nobel Prize for his earlier work with slow neutrons. The connection of nuclear fission with the Manhattan Project of the Second World War will not be considered here.

The introduction during the 1950s of commercially operated nuclear fission reactors marked the beginning of the period of a rapidly expanding usefulness of the nuclear sciences. As growing numbers of radionuclides were produced it became necessary to supplement the periodic table shown in Figure 1.1 with the far more comprehensive Tables of Nuclides.

Tables of Nuclides list the unstable as well as the stable nuclei of atoms, sorted by their proton numbers p (equivalent to Z) (vertical) and their neutron numbers n (horizontal), as well as offering useful explanations of properties of nuclear particles and their interactions. A small, and somewhat simplified section of such a table is shown as Figure 1.3, derived from the *Chart of Nuclides* published by the Kernforschungszentrum Karlsruhe,

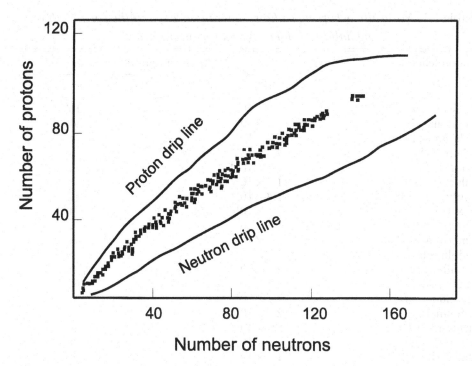

Figure 1.4.   A proton vs. neutron diagram of nuclides showing the stable nuclides as filled squares. The drip lines mark the limits beyond which nuclides cannot accept additional protons or neutrons (based on Hey and Walters (1987, Ch. 5)).

Germany (CoN, 1998). A similar table is *Nuclides and Isotopes*, produced by the General Electric Co. (1989). Other details about nuclear decay data are listed in later chapters.

1.3.4   The role of the neutron-to-proton ratio

Even the small proportion of data included in Figure 1.3 makes it clear that a nuclide cannot be stable unless its ratio of neutrons to protons, n/p, remains within relatively narrow limits. Figure 1.4 shows that it has to be the greater, the greater the Z number.

The ratio n/p for stable nuclides increases from 1.0 to 1.5 as Z or p increases from 2 to 83. The greater the number of protons in a nucleus, the greater the repulsive force that has to be contained, which is done (amongst other things) with the help of neutrons. For elements with $Z > 83$, the repulsive forces between the protons (84 and more) are so large that the nuclei can no longer be kept stable (Figure 1.4). As a result, all nuclides of elements with $Z > 83$ are radionuclides and the elements are known as radioelements. Three chains

Table 1.1. *Naturally occurring radioisotopes in the crust of the earth with very long half lives and at very low concentrations.*

Radionuclide	Approximate values	
	Half life (years)	dps per ton[a]
Potassium-40	1.3×10^9	10^6
Rubidium-87	5×10^{10}	200
Cadmium-113	9×10^{15}	h
Indium-115	4×10^{14}	h
Tellurium-128	1.5×10^{24}	h
Tellurium-130	1×10^{21}	h
Lanthanum-138	1.4×10^{11}	12
Neodynium-144	2×10^{15}	h
Samarium-147	1×10^{11}	h
Samarium-148	8×10^{15}	h
Gadolinium-152	1×10^{14}	h
Hafnium-174	2×10^{15}	h
Lutetium-176	4×10^{10}	60
Tantalum-180	1×10^{13}	h
Osmium-186	2×10^{15}	h
Rhenium-187	5×10^{10}	2
Platinum-190	6×10^{11}	h
Lead-204	1×10^{17}	h

[a] The h signifies < 1 decay per second (dps) per ton. For several radionuclides it is ≪ 1 dps per ton.

of naturally occurring radioisotopes are shown in Figure 1.5, but omitting a few isotopes with intensities below 2%.

A large number of stable elements have between two and six (or more) stable isotopes with their sequence as often as not interrupted by radio-isotopes. Radioisotopes whose neighbours are stable isotopes must have n/p ratios close to the values applying for stable nuclides. This is one, though not the only, reason why radioisotopes have very long half lives (Table 1.1). In Figure 1.3 there is calcium-41, $T_{1/2} = 1 \times 10^5$ years, and potassium-40, the half life of which (1.3×10^9 y) is comparable to the age of the solar system, 4.6×10^9 y.

As a rule, a radionuclide is the more unstable the greater the difference of its n/p ratio from that of the nearest stable isotope of the same element. There are exceptions when half lives do not decrease monotonically, but the general trend is readily verified in Figure 1.3.

Table 1.2. *Products of radioactive decays.*

Decay product	Symbol	Mass (MeV)	Charge (1.6×10^{-19} C)	Comments
Neutrino	ν	0	0	Neutrinos and anti-neutrinos are emitted from atomic nuclei together with respectively positrons and negatrons and also during electron capture decays, when an anti-neutrino is emitted with each electron capture decay
Neutron	n	940	0	Emitted during spontaneously occurring fission of certain heavy nuclides ($Z > 92$)[a]
X and γ rays	X, γ		0	Electromagnetic radiations. Their inertial mass is zero but they have an energy equivalent mass which can be calculated from the Einstein equation
Electron	e^-	0.51	1 unit $-$ve	These three particles are alike in rest mass, but the electron and negatron each have unit negative charge, while the positron has unit positive charge. Negatrons and positrons are emitted from the nuclei of atoms whereas electrons are extranuclear components of atoms
Beta particle or negatron	β^-	0.51	1 unit $-$ve	
or positron	β^+	0.51	1 unit $+$ve	
Alpha particle	α	3730	2 units $+$ve	These particles are nuclei of helium atoms each consisting of two protons and two neutrons

[a] An example is californium-252 ($Z = 98$). Except following spontaneous fissions, neutrons (and protons) are only emitted spontaneously from atomic nuclei as components of α particles.

1.3.5 An introduction to properties of radiations emitted during radioactive decays

Table 1.2 lists radiations emitted during radioactive decays of interest for applications. The neutrino is included, but it will be referred to only rarely in this book.

The table shows the mass and electric charge for each radiation. The mass is expressed in energy units (million electronvolts, MeV) calculated from Einstein's equation. The electric charge is stated as a multiple of the charge

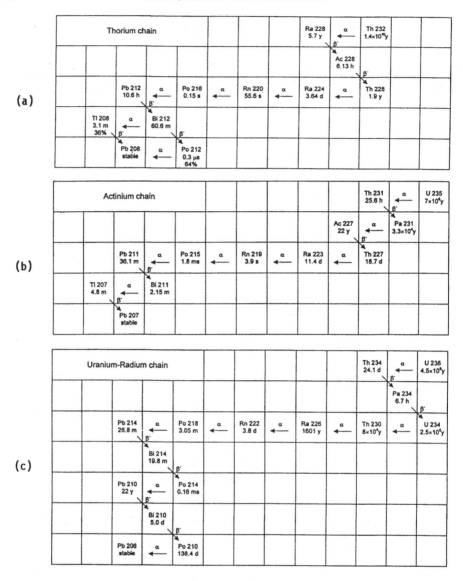

Figure 1.5.   The three naturally occurring radioactive decay chains starting respectively with (a) thorium-232 ($T_{1/2} = 1.6 \times 10^{10}$ years), (b) uranium-235 ($T_{1/2} = 7 \times 10^8$ years) and (c) uranium-238 ($T_{1/2} = 4.5 \times 10^9$ years). A few low percentage decays (<2%) have been omitted.

on the electron (Section 1.2.2). The comments will be supplemented in due course.

Alpha, beta and gamma radiations are commonly grouped together and it is usual to characterise them by differences between their penetrating powers. In fact, differences in penetrating power are due to differences in their mass and electric charge (Table 1.2) which then cause differences in the intensities with which these radiations interact with the atoms of the material through which they travel (see below).

The emission of α or β particles from a nucleus leads to a loss or gain in the number of its protons, a loss when emitting α or β^+ particles but a gain when emitting β^- particles. These changes account for the fact that nuclear transformations signalled by α or β particles lead to daughter nuclides belonging to different chemical elements from the parent. In contrast, the emission of uncharged γ rays from their nuclei takes away energy but leaves their Z number unchanged. An α particle emitting parent of atomic number Z has a daughter of atomic number $Z - 2$. If a β^- particle emitting parent has atomic number Z, its daughter has atomic number $Z + 1$ (see Figure 1.5 for examples).

A beam of electrically uncharged 200 keV γ rays, a low energy for these rays, could travel through 5 mm thick steel losing no more than half its initial intensity. For 1 MeV γ rays this distance is 15 mm. In contrast, a beam of 1 MeV β particles is completely absorbed in 0.6 mm steel.

Gamma rays interact almost invariably with atomic electrons, knocking them out of their atoms in relatively few, randomly spaced collisions, passing vast numbers of atoms without interacting. In contrast, electrically charged α or β particles interact with atomic electrons via continuously acting electric (Coulomb) forces between the particles and outer atomic electrons of which there are tens of millions per millimetre of a solid, so accounting for the very much stronger interactions of charged particles with the atoms through which they travel than applies to electromagnetic radiations.

Although each ionisation or excitation takes, on average, only a small number of electronvolts, a 1 MeV β particle loses all its energy after travelling through no more than the 0.6 mm steel as mentioned earlier. For the far more intensely interacting α particles, the range in steel is less than about 0.003 mm/MeV. Uncharged elementary particles such as neutrons are, however, in a different category. A brief introduction to neutron sources and how they are used will follow in the next section.

1.3.6 Another nuclear radiation: the neutron

Neutrons have many important applications. Being electrically uncharged like γ rays, neutrons too are penetrating radiations, though there are exceptions. The emission of α or β or γ radiations from atomic nuclei occurs spontaneously. Neutrons are emitted spontaneously only during spontaneous fissions which occur at a very low rate in a few high atomic number elements.

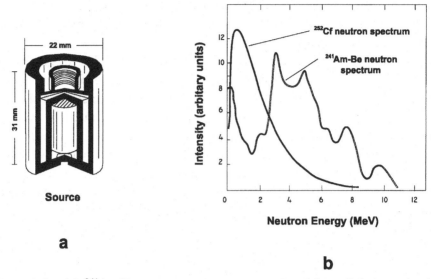

a

b

Figure 1.6. (a) ^{241}Am/Be neutron source encapsulated in stainless steel tubing; dimensions in millimetres. (b) The energy spectra for neutrons from ^{241}Am/Be and from spontaneous fissions of californium-252 ($T_{1/2} \sim 2.6$ y, $f_\alpha \sim 3.1\%$, adapted from IAEA, 1993, Ch. 5, Figure 5.1).

A rare case where spontaneous fissions account for some 3% of the decays is the trans-uranium element californium-252 ($Z = 98$, $T_{1/2} = 2.64$ y). The other ~97% are α particle decays. Californium-252 can be prepared as a portable neutron source (Figure 1.6(a)). The spectrum of its neutrons covers a range similar to that due to other neutron sources but differs in shape (Figure 1.6(b)).

High-intensity, high-energy fluxes of neutrons for scientific and industrial applications and, in particular, for the modification of the atomic properties of materials are produced in research reactors (Section 1.4.1) and also in so-called spallation sources, which are much more powerful (and much more expensive) than research reactors. Another nuclear reaction yielding neutrons is the bombardment of tritium atoms by accelerated deuterons (^{2}H) producing 14 MeV neutrons. The tritium bombardment can be effected in relatively small and compact accelerators which can be used during borehole logging applications (Section 7.4.2, see also Cierjacks, 1983).

Neutrons have many applications 'in the field' where they are emitted from portable sources also known as isotopic sources (Figure 1.6(a)). Neutrons are commonly produced using the 5.5 MeV α particles from americium-241 (Section 3.2.1) interacting with beryllium ($Z = 4$). The reaction is ^{9}Be + ^{4}He → ^{12}C + n, with each neutron followed by a 5.7 MeV γ ray.

The rate at which neutrons are generated in portable $\alpha \rightarrow n$ sources is rarely greater than 10^6 neutrons per second, a limit set in part to ensure radiation safety. This is a relatively low emission rate, but easy portability is a sufficiently compensating advantage. Another advantage is the fact that source and detector can be mounted together because proportional counters used for this purpose are efficient detectors for neutrons once they have been slowed to room temperature energies (0.025 eV) during applications, but the counters ignore fast neutrons from the source (Section 5.4.4). Compactness is here particularly useful because neutrons are often required in confined spaces, e.g. in narrow boreholes when exploring for hydrocarbons or water.

Doses to experimenters using these $\alpha \rightarrow n$ reactions are kept small because the 59.5 keV γ rays from ^{241}Am are largely absorbed in the metal wall of the neutron source (Figure 1.6(a)) while the 5.7 MeV γ rays, along with the neutrons, are emitted at a sufficiently low rate to make it easy to prevent effects harmful to health.

As predicted by quantum theory, neutrons are not only behaving like particles but also like waves. Collimated beams of neutrons from research reactors can be diffracted in organic or inorganic crystals just like beams of X rays, though X rays interact preferentially with high atomic number materials whereas neutrons interact preferentially with low atomic number materials, especially hydrogenous materials (Section 7.4.3).

Neutron detection will be introduced in Section 5.4.4 together with detection methods for other nuclear radiations. It is first necessary to gain some familiarity with the role of neutrons in nuclear research reactors.

1.4 Activation processes

1.4.1 Nuclear fission reactors

Nuclear power production could make impressive advances on the world stage because it was supported by extensive research based on relatively low-powered research reactors (Section 1.1.3), the power output being rarely as much as 30 MW whereas it is rarely less than 600 MW for power reactors.

Research reactors are designed to optimise conditions for the production of high fluxes of neutrons which are used to investigate the properties of numerous types of materials. Neutrons are also employed for the production of radionuclides to be used in science, medicine and industry. Today's research reactors employ a large range of operating procedures which continue to be improved upon, particularly as regards safe operating conditions. The following description of their operation should be read bearing this in mind.

The reactor fuel, commonly uranium oxide, is enclosed in metal tubes that are kept tightly sealed at all times to prevent the escape of the highly radioactive fission products. The metal tubes are immersed in moderating material, normally water or heavy water, for reasons to be stated presently. The about 10 cm diameter hollow interior of these tubes holds the material to be activated or irradiated.

When the research reactor is operating, its core is bathed in a gas of neutrons which collide randomly with uranium atoms. When neutrons interact with atoms of the isotope uranium-235 (^{235}U), they are likely to cause them to break up (fission) into lighter atoms while also emitting on average about 2.4 neutrons per fission though only one of these is allowed to keep the fission process going (see below). The fission products are strongly radioactive and of very high kinetic energy. In power reactors the fission energy is used to operate large electric power stations. In the much smaller research reactors the liberated energy is commonly only a few megawatts and so is insufficient to be useful.

In order to maximise the neutron flux, the uranium in the fuel elements is enriched in its U-235 isotope, the concentration of which is only about 0.7% in natural uranium. For many power reactor designs the enrichment level is about 3%. For research reactors one normally requires uranium with at least 20% enrichment to obtain a satisfactorily high neutron flux. Reactors are surrounded by thick concrete shielding to protect operators from the neutron flux and the high-energy γ rays that are generated in the core following promptly on from neutron interactions. The shielding contains access ports for the loading of material which is to be activated in the 0.1 m diameter hollow cores of the fuel elements. The ports also serve for neutrons to be withdrawn from the core for a variety of applications.

The energies of the neutrons released during fissions range upwards to about 8 MeV with an average near 2 MeV. Fast neutrons are not nearly as efficient in fissioning U-235 nuclei as are thermal neutrons, so-called because their average energy is similar to that of gas molecules at the prevailing temperature, e.g. near room temperature, when the average neutron energy is about 0.025 eV. It is not only the fission process which is kept going most efficiently by thermal (slow) neutrons. Stable nuclides are also more efficiently activated with slow than with fast neutrons (Section 1.3.3), though fast neutron activations (neutrons at thousand or million electronvolt energies), have their own importance.

To reduce (moderate) the energy of the initially fast neutrons, the uranium fuel is mixed with or surrounded by moderating material, commonly heavy water (D_2O). Collisions with hydrogen and deuterium nuclei of the moderating material quickly reduce the energy of the large majority of the fast neutrons as required. Efficient neutron absorbers which can be quickly lowered into or raised from the reactor core ensure that the fission process proceeds at a strictly uniform rate.

1.4.2   Thermal neutron activations

Thermal neutron activations are of greatest interest here and will be discussed presently. However, there are activations using neutrons of thousand and million electronvolt (fast) energies that also have important applications as will be seen in Section 7.4. A few words about fast neutron activations will follow in Section 5.4.4.

The efficiency of thermal neutron activations of stable nuclides at a given neutron flux density (neutrons per square centimetre per second depends on the probability with which the nuclides of a given isotope capture thermal neutrons. In Figure 1.3 this probability, known as the thermal neutron capture cross section per atom for the reaction, is labelled by the Greek letter sigma (σ). The unit for the cross section is the barn (10^{-24} cm^2), designating an area approximately equal to that of the cross section of the nucleus of an atom.

Reactor activations for the production of radionuclides are carried out in a constant flux of neutrons and so at a constant rate. This is so because in the large majority of irradiations no more than a tiny fraction of the irradiated material ($<10^{-4}$ %) is activated. Hence, the neutron flux density and also the quantity of stable nuclides remain practically unchanged, and so does the rate of activations.

The thermal neutron capture cross sections in Figure 1.3 are shown only for stable nuclides (black squares). Radionuclides (white squares) also absorb thermal neutrons though mostly less efficiently. In each case the thermal activation cross section per atom of a nuclide measures the probability with which individual nuclides exposed to the neutron flux capture thermal neutrons. Each capture increases the mass number of the absorbing nucleus by one unit, which, more often than not, causes originally stable nuclides to become radioactive (Figure 1.3).

Neutron captures, thermal or fast, are almost always followed promptly by highly energetic γ rays and so are known as (n,γ) reactions. These reactions are not only responsible for the production of radioisotopes but also, for example, for neutron activation analysis to be described in Section 7.4.4.

1.4.3 Activation and decay

A typical example of an (n,γ) activation is the reaction:

$$^{31}P \ (n,\gamma) \ ^{32}P. \tag{1.1}$$

Adding a neutron to the stable ^{31}P nucleus forms ^{32}P, which is radioactive. Radionuclides begin to decay as soon as they are formed so that activation and decay always proceed together. The decay of ^{32}P can be written

$$^{32}P \ (\beta^-, \ T_{1/2} = 14.3 \ \text{days}) \rightarrow \ ^{32}S \ (\text{stable}). \tag{1.2}$$

The emission of a β^- particle from the nucleus of phosphorus-32 leaves its mass number unchanged at $A = 32$, but it results in the conversion of a neutron into a proton, causing the Z number of the nucleus to change from 15 to 16, so becoming sulphur-32, which is stable (Figure 1.3).

It is a fundamental law governing decay rates in samples of radionuclides that the number of radioactive atoms (dN) decaying in time (dt) is proportional to the total number of these nuclides (N) present at that time (see Eq. (1.5) in Section 1.6.2). When an activation begins many more atoms are activated than decay causing the decay rate in the sample to increase with irradiation time until the number of decaying atoms approaches the number that is being activated when the activation approaches saturation.

The approach to saturation is asymptotic. As a rule, the activated material is removed from the reactor and processed for dispatch to users before its activity is too close to saturation. This is done because the longer a material is irradiated, the more likely the generation of radionuclidic impurities, while the residual gain in activity is only very small.

1.4.4 Other activation processes

The production of neutron-poor radionuclides

Tables of Nuclides show that for every element with $Z > 3$ the stable nuclides have radioactive neighbours on either side, the neutron-rich radionuclides on the right and the neutron poor on the left. Neutron-rich radionuclides normally decay by β^- particle emissions while neutron-poor radionuclides decay by β^+ emissions or, more frequently, by electron capture (EC) to be discussed in Section 3.6.1.

To generate neutron-poor radionuclides from stable nuclei the latter have to accept positively charged elementary particles, commonly protons which can only penetrate into positively charged nuclei when they carry enough kinetic energy to overcome the repulsive forces between equally charged particles.

Protons ($Z = 1$) require several million electronvolts of kinetic energy to enter even low Z number nuclei. Alpha particles ($Z = 2$) require considerably higher energies. The following are typical reactions initiated by protons or α particles: (p,n), (p,4n), (α,n), (α,2n). The first letter names the captured particle; the second letter names the particle or particles emitted immediately following the capture. A (p,n) or (α,n) reaction leads to the loss of a neutron from the target nucleus, so decreasing n/p as required.

These reactions are effected by high-energy particle accelerators such as cyclotrons which transfer sufficient energy to protons or α particles to permit them to penetrate into positively charged nuclei as required. Alpha particles emitted from radionuclides can also have sufficient energy to cause (α,n) reactions in low Z number nuclides such as beryllium ($Z = 4$), which permits the production of portable neutron sources that are very useful in a number of applications (Sections 1.3.6 and 7.4.2).

Positron emitters for nuclear medicine

Neutron-poor radionuclides decay by electron capture or with the emission of positrons. At present it is practitioners of nuclear medicine who are among the principal users of positron emitters for a procedure known as positron emission tomography (PET).

When destined for medical use positron emitters are produced in cyclotrons located in the grounds of a hospital. This is done because of the very short half lives of the positron emitters of greatest clinical interest: carbon-11 ($T_{1/2} = 20$m), nitrogen-13 ($T_{1/2} = 10$m), oxygen-15 ($T_{1/2} = 2$m) and fluorine-18 ($T_{1/2} = 110$m). The clinical interest arises in part from the fact that the nuclides, except fluorine-18, are constituents of organic molecules participating in body metabolism. When activated, these short-lived radionuclides are incorporated into pharmaceutical substances as radiotracers for investigations of cancers and of metabolic malfunction, especially in the brain. The fourth element normally present in metabolic reactions is hydrogen. Fluorine-18 has proved itself an effective replacement for hydrogen which has no suitable radioisotope.

Carbon-11, as well as ^{13}N, ^{15}O and ^{18}F, can be produced by relatively low energy proton beams (~ 6 MeV). Other neutron-poor, medically important radionuclides, e.g. gallium-67 ($T_{1/2} = 3.26$ d), iodine-123 ($T_{1/2} = 13.21$ h) or thallium-201 ($T_{1/2} = 3.04$ d), are produced with higher energy proton beams, up to around 30 MeV. Having sufficiently long half lives, they can then be distributed to the nuclear medicine departments which require them.

Short-lived positron emitters are immediately synthesised into pharmaceutical substances selected for their clinical requirements and transferred as quickly as possible from the cyclotron to the nuclear medicine department where they can be administered to patients within minutes from the time of their production, if necessary.

Sufficient activity is injected into patients for positron–negatron (β^+–β^-) annihilations to occur at a rate of 10^4 to 10^5 events/s. At that rate the detection of pairs of coincident 511 keV γ rays produced by these annihilations (Section 3.3.1) yields enough information for the fully computerised signal processing system to produce well resolved images of the biochemical processes in the bodies of patients.

Positron emitters to effect PET are also increasingly used in industry, so adding to their importance.

1.5 Short and long half lives and their uses

1.5.1 Generators for short half life radionuclides

The data in Figure 1.3 show that it is always possible to produce, for each element, radioisotopes with half lives of a few hours or less. Short half life radionuclides are the preferred choice for applications (Section 7.2.1), not least because there are then few problems with the disposal of radioactive residues. However, short-lived radionuclides have to be used as soon as they have been produced, so causing logistic problems unless there is direct access to a nuclear reactor or an accelerator.

Fortunately, there are a few long-lived radioactive parents followed by short-lived radioactive daughters which can be eluted from the parent independently of location in so-called radionuclide generators. These generators have two convenient characteristics: (i) a radionuclide with the half life of a few days or less is continuously regenerated as the daughter of a much longer-lived parent (Section 1.6.3); (ii) since parent and daughter belong to different chemical elements the radioactive daughter can be separated from the parent by a simple chemical procedure. The terms parent and daughter are customarily used to designate respectively any radionuclide and its decay product which may or may not be active.

The simplest version of a radionuclide generator is a glass tube, some 15 mm diameter and about 50 mm long, fitted at one end with a glass stopcock and, if necessary, protected by a lead shield. The tube is filled with an ion exchange resin selected to absorb the parent but not the daughter. When washed with a readily available solvent, the long-lived parent remains firmly attached to the resin while the short-lived radioactive daughter is washed out, when the eluted solution is available for applications.

Following an elution, the radioactive daughter at once begins to regrow into the parent nuclide until its decay rate reaches equality with the rate at which it is generated although elutions could of course be made before the daughter reaches equilibrium with the parent.

This account of radionuclide generators and their operation is greatly simplified. In practice one requires numerous precautions to avoid errors. Radionuclide generators are being supplied commercially with detailed instructions on how to operate them safely and economically. A long-lived parent–short-lived daughter pair which is routinely separated in commercially available generators is strontium-90 (^{90}Sr, $T_{1/2} = 28.5$ y) → yttrium-90 (^{90}Y, $T_{1/2} = 64$ h) → zirconium-90 (^{90}Zr, stable). The radioactive daughter, ^{90}Y, is a pure β particle emitter with its high endpoint energy of 2.28 MeV making it useful for many applications.

A list of radionuclides of moderate to long half lives (including ^{90}Sr), followed by short half life daughters (from fractions of a minute to a few days), suitable for use in radionuclide generators is given by Charlton (1986, Section 4.6). Parent–daughter relationships will be taken up in more detail in Section 1.6.

1.5.2 Isomeric decays with applications to nuclear medicine

Only a minority of radioactive transformations proceeds to the ground state of the daughter. The majority proceeds to excited states when de-excitations could occur subject to a conveniently long half life just like nuclear transformations. If so, the de-excitation energy is commonly emitted as γ rays which are then available for applications.

For de-excitations with sufficiently long half lives (this could be a few seconds for highly experienced operators but normally one requires a few minutes), the excited, or metastable (m) nuclide, can be used just like a daughter radionuclide, when the decay is known as an isomeric transition (IT).

For example, ^{137m}Ba, the isomeric daughter of caesium-137 (see Figure 3.4(c) and Table 4.1) de-excites to the ground state of ^{137}Ba with the emission of the 662 keV γ ray and a half life of 2.5 minutes. It is normally eluted from a radionuclide generator charged with ^{137}Cs-chloride using 0.05 M HCl. The γ rays are then available for applications.

A medically important isomeric decay is that of technetium-99m ($T_{1/2} = 6.01$ h, $E_\gamma = 140$ keV, Figure 4.7), the isomeric daughter of molybdenum-99 ($T_{1/2} = 66.0$ h). The ^{99m}Tc is eluted from commercially available generators containing its parent ^{99}Mo which, with its 66.0 h half life, remains capable of producing what can be a highly pure product for about a week following its activation. ^{99m}Tc is the most extensively used radiotracer in nuclear medicine (itself one of the largest users of radionuclides), not least because it combines firmly with numerous pharmaceutical substances used for the investigation of a large range of diseases. Its half life is long enough to conveniently permit investigations of patients using γ cameras, but short enough to minimise problems with radioactivity absorbed by patients. Its 140 keV γ ray energy is extremely well suited for obtaining clear, well resolved radiographs from patients of all ages.

1.5.3 Radionuclides with very long half lives

A relatively large number of radionuclides with $Z < 84$ decay to stable daughters with half lives of a million years and longer. It could be useful to know that they are available, though rarely at specific activities exceeding a few million becquerels per gram.

Very long-lived radionuclides ($T_{1/2} > 10^6$ years) can be divided into two groups. Radionuclides in the first group were just referred to. They account for rarely more than a very small fraction of the element, e.g. 0.012% for potassium-40 (Figure 1.3). They are shown in Table 1.1, though this list is not complete. The second group includes the long-lived members of the three decay chains shown in Figure 1.5.

The very long half lives measured for thorium-232 ($Z = 90$) and uranium-238 ($Z = 92$) represent maxima for radionuclides with Z numbers exceeding 83. When nuclides with $Z > 92$ were synthesised (Figure 1.1), it was found that half lives decrease as Z increases. This is not surprising since sufficiently long-lived radioelements would still be with us. A large number of high atomic number elements ($Z > 92$) have now been synthesised, as shown in Figure 1.1.

By the mid-1990s all elements up to $Z = 109$ had been named after the researchers who had been responsible for their discovery. Element 109 (at. wt 268) was named after Professor Lise Meitner (meitnerium) who had made decisive contributions to the discovery of nuclear fission. These high atomic number elements are highly unstable with half lives of a few milliseconds or only small fractions of a millisecond. However, at still higher Z numbers there could be 'islands of stability' with isotopes returning to longer half lives. Details are here out of place.

Referring back to Table 1.1, the half lives of several of the radionuclides listed are very much longer than the half lives of thorium-232 or uranium-238. This is no doubt due, at least in part, to their n/p ratios that are in the range where other isotopes of the same element are stable.

1.5.4 The energetics of decays by alpha and beta particle emissions

The origin of the energies liberated during nuclear decays can be explained in terms of the mass–energy relationship which is a consequence of the special relativity theory proposed by Einstein in 1905. The relevance of that theory was not realised until well into the 1920s.

The fact that ^{32}P undergoes radioactive decay to ^{32}S with the emission of β^- particles (Section 1.4.3) happens for several reasons, two of which are relevant here: (i) the n/p ratio of ^{32}P is larger than permitted for stability; (ii) the nuclear masses of ^{32}P and ^{32}S differ by an amount which can provide the required energy. A difference in nuclear mass is not by itself a sufficient reason for causing nuclear instability. If that were so all nuclides would be unstable since no two neighbours have exactly the same nuclear mass.

Let the masses (in amu, see Section 1.3.2) of an unstable nuclide (the parent), and its stable daughter be respectively m_P and m_D. To a first approximation, the conditions providing the energy for α or β particle decays are:

$$\text{for } \alpha \text{ decay:} \qquad m_P > (m_D + m_\alpha), \qquad (1.3)$$

$$\text{for } \beta \text{ decay:} \qquad m_P > (m_P + m_\beta), \qquad (1.4)$$

where m_α and m_β stand for the masses of the α and β particles respectively, also in amu. Expressions (1.3) and (1.4) were derived subject to a number of simplifications, but that does not affect the argument.

Applying expression (1.4) to the decay $^{32}P \rightarrow ^{32}S$ (Eq. (1.2)), one obtains from tables of atomic masses that $m_P = 31.9739$ amu, $m_D = 31.9721$ amu with the mass of the emitted β particles equal to 0.0005 amu (see e.g. Mann *et al.*, 1980, Ch. 5). Allowing for some rounding, the parent has a mass excess of 0.0018 amu, equivalent to an energy excess of 0.0018×932 MeV $= 1.7$ MeV, approximately, in satisfactory agreement with the observed maximum energy of the β particles emitted by ^{32}P (Figure 1.3). The mass of the neutrino is assumed to be zero. Neutrinos are defined in Table 1.2 and their role in β particle decay will be briefly referred to in Section 3.3.1.

Alpha particles are over 7300 times more massive than β particles (Table 1.2). It is thus not surprising that only relatively heavy radionuclides can undergo α particle decay. These decays are commonly too complex for the

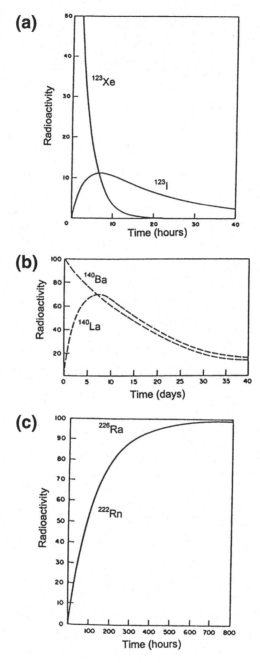

Figure 1.7. Three radioactive parent → radioactive daughter decays. (a) The growth and decay of iodine-123 ($T_{1/2} = 13.2$ h) which survives its parent xenon-123 ($T_{1/2} = 2.08$ h). (b) Lanthanum-140 ($T_{1/2} = 40.27$ h), growing to transient equilibrium with its parent barium-140 ($T_{1/2} = 306.5$ h). (c) Radon-222 ($T_{1/2} = 3.82$ d) growing to secular equilibrium with its parent radium-226 ($T_{1/2} = 1601$ y). During the measured period (800 h) ^{226}Ra remains effectively unchanged.

purpose of this book. However, it can be shown that the energy equivalent of the mass difference agrees with the measured kinetic energy of the α particles as well as can be expected from the uncertainties affecting these measurements.

1.6 Parent half lives and daughter half lives

1.6.1 Three cases

All radionuclides are parents but not all daughters are radioactive daughters. The present concern is with radionuclides that decay to radioactive daughters.

One can distinguish between three cases, illustrated in Figure 1.7:

(a) $(T_{1/2})_P/(T_{1/2})_D < 1$
(b) $(T_{1/2})_P/(T_{1/2})_D > 1$ but $< 10^3$,
(c) $(T_{1/2})_P/(T_{1/2})_D > 10^3$.

Case (a) leads to the daughter nuclide surviving its parent. It is used for the production of radionuclides but will not be further considered here. Cases (b) and (c) will be dealt with after discussing equations describing radioactive parent → radioactive daughter decays which are used to calculate the results of frequently employed procedures. For details about the derivation of these equations readers are referred to Mann *et al.* (1980, Ch. 2).

1.6.2 Decay chain calculations

The decay rate $(-dN/dt)$ of atoms in samples of radioactive materials is directly proportional to the number (N) of radioactive atoms present in the sample (Section 1.4.3). In symbols:

$$-dN/dt \propto N = \lambda N, \tag{1.5}$$

where λ is known as the decay constant. No two radionuclides have the same value of λ, which means that no two have the same half life (Eq. (1.7)). The minus sign signifies that the number of radioactive atoms is decreasing as the time t is increasing.

Re-arranging the terms and integrating leads to the radioactive decay equation:

$$N_t = N_0 \cdot \exp(-\lambda \times dt), \tag{1.6}$$

where N_t stands for the number of radioactive atoms at time t, N_0 for the number at some reference time t_0 and $dt = t - t_0$. The decay constant λ is related to $T_{1/2}$, the half life of the radionuclide by:

$$T_{1/2} = \ln 2/\lambda = 0.693/\lambda. \qquad (1.7)$$

Applying these relationships to the decay involving a radioactive parent → radioactive daughter pair, the number of atoms of the parent N_P and that of the radioactive daughter N_D can be shown to be related by:

$$N_D = f N_P [\lambda_P/(\lambda_D - \lambda_P)] \times [\exp(-\lambda_P t) - \exp(-\lambda_D t)], \qquad (1.8)$$

where t denotes the time since the preceding elution washed out all radioactive daughter atoms and f is the branching ratio for the parent decays. For example, for the decay to the excited state of the daughter $^{137}Cs \rightarrow {}^{137m}Ba$ one has $f = 0.95$, the balance of ^{137}Cs decays going direct to the ground state of ^{137}Ba (Figure 3.4(c)).

1.6.3 Transient and secular equilibrium

Diagrams (b) and (c) in Figure 1.7 show examples of transient and secular equilibria respectively. Case (b) decays lead to transient equilibrium. This is so when the half lives of parent and daughter and so the difference between these half lives are of the order of weeks or only hours. Whenever the conditions for transient equilibrium apply the daughter activity goes through a maximum as shown (Figure 1.7(b)) and then enters the transient equilibrium phase when A_D decays at the same rate as A_P but with A_D exceeding A_P.

On simplifying Eq. (1.8), using the conditions applying to case (b), it is seen that when t is long enough for parent and daughter decays to reach equilibrium, one obtains:

$$A_D/A_P = f \times (T_{1/2})_P/[(T_{1/2})_P - (T_{1/2})_D]. \qquad (1.9)$$

The magnitudes of A_P and A_D prior to reaching equilibrium have to be measured or have to be calculated from Eq. (1.8).

For the parent–daughter decay Ba-140 → La-140 (Figure 1.7(b)), one has $f = 1$, $(T_{1/2})_P = 306.2$ h and $(T_{1/2})_D = 40.3$ h. Applying Eq. (1.9), on reaching equilibrium, one obtains $A_D = 1.15 A_P$, so that the daughter settles down to decay in transient equilibrium with the parent, but its activity remains 15% above that of A_P.

Case (c) applies when $(T_{1/2})_P \gg (T_{1/2})_D$. Parents in that category commonly have very long half lives and the ratios $(T_{1/2})_P/(T_{1/2})_D$ are of order 10^4 or greater (see below). This means that, in practice, A_P remains effectively constant while A_D builds up from zero (following an elution) to its saturation value. Using the relationship between N_D and N_P as defined in Eq. (1.8) and replacing λ by $T_{1/2}$ (Eq. (1.7)), Eq. (1.9) can be made to take the simple form:

$$A_D = f \times A_P \left[1 - \exp[-\ln 2 \times t/(T_{1/2})_D]\right], \qquad (1.10)$$

where t is again the time since the last elution of the daughter. When t equals no more than five of the short daughter half lives (they are at most a few days), one has A_D equal to A_P within 3.1% and when t equals ten daughter half lives one has A_D reaching equality with A_P within 0.1%, with A_D continuing to approach the effectively constant A_P asymptotically as expected from Eq. (1.10). This is indicated in Figure 1.7(c) where the ratio $(T_{1/2})_P / (T_{1/2})_D$ is no less than 1.53×10^5.

If a chain of naturally occurring radionuclides (Figure 1.5) is in secular equilibrium, all daughters decay at the rate equal to that of the first parent. While these chains could date back to the time of the solidification of the earth, it is more likely that they would have been frequently fragmented by physical or chemical forces when the fragments that had separated from the first parents decay away, while the part of the chain connected to the first parents grows to replace what has been lost, except for decay of the parent. The latter case is the basis of radioactive dating techniques employing primordial radionuclides (Section 9.4.1).

Chapter 2

Units and standards for radioactivity and radiation dosimetry and rules for radiation protection

2.1 Introduction

Chapter 1 introduced basic properties of radioactivity, including radio-activation processes and criteria for nuclear stability. Two other subjects necessary when applying radioactivity will be introduced in this chapter: units and standards employed for radioactivity measurements and radiation dosimetry, and an introduction to rules and procedures developed to permit radiation applications to be carried out safely and efficiently with adequate protection from potentially harmful effects due to ionising radiations.

Appendix 2 at the end of this book lists publications carrying comprehensive information on health physics and radiation protection; these could be consulted if and when in doubt about safe operating conditions. The comments in this chapter and elsewhere in this book can cover no more than a small part of a large subject.

2.2 Units and standards of radioactivity

2.2.1 *A summary of their characteristics*

The consistency of results of physical measurements made in different locations depends on the consistency of the units employed for this purpose. All industrialised countries have passed laws accepting an international system of metric units for physical measurements known as Système International (SI) (BIPM, 1985) along with provisions for monitoring the coherence of these units by standards established in national and international standards laboratories. These laboratories cooperate around the globe to ensure that the units employed in their respective countries (see Section 2.3.2 below) are identical in size with those used in other countries, an essential activity because units for physical measurements have high industrial and commercial as well as scientific importance.

The SI unit for the measurement of the activity of a radionuclide is the becquerel. It has to be used with care because the decay of all radionuclides is measured in becquerels even though no two of them decay with the emission of the same mix of radiations as regards their types, energies and intensities. (For collections of decay data, see e.g. NCRP (1985, Appendix A3) or Chapter 8 of *The Radiochemical Manual* (Longworth, 1998) or National Institute of Science and Technology (NIST) data available over the Internet (see Appendix 3, Table A3.1). For other references containing collections of decay data see Section 5.5.1 in this book.)

The fact that no two radionuclides have the same decay characteristics makes it necessary that an accurate verification of the decay rate of a radionuclide R has to be made using a standard which is a sample of R. No other radionuclide will do unless one accepts uncertainties that would be the larger the less similar the decay schemes.

There are two other characteristics that are specific for units and standards of radioactivity. First, there is the randomness of decay events: equal intervals of time do not contain exactly equal numbers of decays except by chance. This important qualification affects all measurements of radioactivity. It will be dealt with in more detail in Sections 6.5.1 and 6.5.2. Second, radioactivity diminishes with time at a rate determined by the half life of each radionuclide (Section 1.3.2). For a number of radionuclides the half life exceeds 10^4 years when the decay rate is imperceptibly small even over decades. Many others decay at a rate such that calibrated samples cease to be useful within days or hours after they were prepared (see e.g. Figure 1.5).

Decay rates measured in becquerels (Eq. (2.1)) refer not only to nuclear transformations such as that from ^{32}P to ^{32}S (Section 1.4.3), but also to other types of decays to be dealt with in later chapters and briefly noted below:

Isomeric decays (Section 1.5.2), when the daughter of a radionuclide created in an excited state de-excites to its ground state subject to a readily measurable half life, e.g. the decay of ^{137m}Ba to ^{137}Ba when $T_{1/2} = 2.55$ m (Figure 3.4(c)).

Parent nuclides decaying to radioactive daughters, e.g. ^{90}Sr decaying to ^{90}Y (Section 1.5.1), when a stated activity, e.g. 10 kBq applies only to the parent, i.e. ^{90}Sr. The activity of the daughter, ^{90}Y, is disregarded unless otherwise stated, e.g. 10 kBq (^{90}Sr + ^{90}Y).

Numerous radionuclides follow two or even three decay paths. An example of the latter is copper-64 (^{64}Cu, $Z = 29$, $T_{1/2} = 12.8$ hours, Figure 3.13(b)). A hundred decays of ^{64}Cu are made up, on average and in rounded figures, of 38 β^- decays (to zinc-64, $Z = 30$), 18 β^+ decays and 44 electron capture decays (both to nickel-64,

$Z = 28$). In cases such as this, an activity A Bq of ^{64}Cu is the sum of the parallel modes. To obtain, say, the positron emission rate in becquerel, one has to multiply A by the positron branching ratio, 0.18.

2.2.2 *The curie and the becquerel*

Initially it was assumed that naturally occurring radioactive substances are radioactive elements (Figure 1.5) whose mass could also serve as a measure of their radioactivity. In early radioactivity research it was decided that the unit of radioactivity should be called the curie and should equal the decay rate of one gram of pure radium, 3.65×10^{10} s^{-1} (Geiger and Werner, 1924). The activity of a pure radionuclide is indeed proportional to its mass, but this fact was soon found to be impractical for use in radioactivity measurements because of the complex parent–daughter inter-connectedness of naturally occurring radionuclides (Figure 1.5) which were the only ones then known and expected to exist.

Eventually the curie (Ci) was re-defined independently of the properties of radium or any other radionuclide as: 1 Ci $= 3.7 \times 10^{10}$ radioactive transformations per second (exactly) .

It was also recognised that radioactivity measurements could not be absolute. The activity of a source cannot be derived from properties of the radionuclide (e.g. its mass). Activity has to be determined by counting the number of nuclear transformations or decays or disintegrations per second (these terms are used interchangeably) in a sample of the radionuclide of interest, such counts being now known as direct radioactivity measurements. Measurements using calibrated instruments or other non-direct ways are known as secondary or relative measurements.

With the introduction of SI units during the mid-1970s the curie had to be replaced by a unit consistent with the new system. This was named the becquerel (Bq) after the Frenchman H. Becquerel who had discovered radio-activity in 1896. However, the curie remains in use, especially in medicine and for many industrial applications, and it will also be used occasionally in this book.

A definition of the becquerel which is sufficient for our purposes is:

$$1 \text{ Bq} \equiv 1 \text{ nuclear decay or transformation per second.} \qquad (2.1)$$

One becquerel denotes a small activity. For this reason the activity of radioactive sources is commonly expressed in multiples of the becquerel, e.g. kilobecquerels (1 kBq $= 10^3$ Bq) or megabecquerels (1 MBq $= 10^6$ Bq). In contrast, the curie is equal to 3.7×10^{10} Bq $= 3.7 \times 10^7$ kBq or 3.7×10^4 MBq, a large activity. The activities of sources are commonly no more than a few millicuries (1 mCi = 10^{-3} Ci) or a fraction of a microcurie (1 μCi = 10^{-6} Ci = 37 kBq).

2.2.3 Secondary standards and secondary standard instruments

Standards of radioactivity are supplied by national and international standard laboratories and authorised commercial laboratories, commonly as a few grams of aqueous solution of a stated radionuclide with a certified radioactivity concentration (Bq/g) and other characteristics, notably a stated level of radionuclidic purity (see also Section 2.3.1). Users of radionuclides should apply to the authority responsible for ionising radiation in their country of residence to obtain information about the supply of radioactivity standards, bearing in mind that whenever possible they should be samples of the radionuclide of interest (Section 2.2.1) or, at least, have closely similar decay schemes.

When radionuclides are used on a larger scale, it is often helpful to employ standard sources to calibrate the response of an ionisation chamber or of a NaI(Tl) detector which is then used as a secondary standard instrument (SSI). The calibration of these instruments will be discussed in Sections 6.3.3 and 6.3.4. The use of SSIs carries two important advantages. First, once calibrated, these instruments can serve for radioactivity measurements (often for long periods of time) while the activities of many of the standard sources used for the calibration have long decayed away.

Second, SSIs are chosen to ensure that their response is a smooth, close-to-linear function of the γ ray energy, though this can commonly only be realised at energies above about 150 keV. Figure 2.1 shows a typical calibration graph of a pressurised ionisation chamber of the type to be discussed later in this book (Sections 5.3.3 and 6.3.3). However, SSIs can only respond to radionuclides that emit sufficiently energetic and sufficiently intense X or γ rays. There are, so far, no SSIs for pure or nearly pure β particle emitters.

Although pressurised SSIs are highly stable instruments, it is necessary to verify the stability of their calibration at least once per year, which should be done at three or four strategically chosen energies along their range. Stability is important since these instruments serve to verify not only the disintegration rates of sources but also the reproducibility of ionisation current measurements. This can be readily done, using a high quality instrument, to within ±1.0% and given enough experience to within ±0.1%.

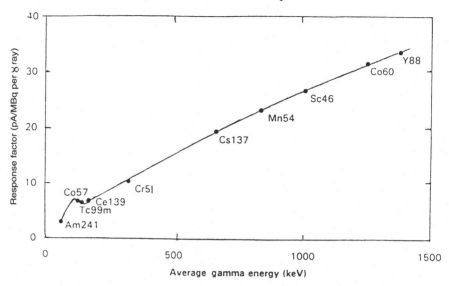

Figure 2.1. Calibration graph of a $4\pi\gamma$ pressurised ionisation chamber. The detection medium is argon at about 2 MPa (~ 20 atm). The ionisation current (pA/MBq) is shown per γ ray which has to be allowed for when the γ ray emission rates differ from unity which is the rule, being e.g. 2.0 for ^{60}Co.

2.2.4 In-house standards

As mentioned above, standard sources are calibrated in terms of national standards. In-house standards are also used. They are sources of long-lived radionuclides which, when measured in an accurately reproducible source-detector geometry, are known to yield accurately reproducible countrates except for necessary corrections. Such sources are useful for monitoring the response of detectors and also provide reference standards for applications.

Radionuclide sources to be used as in-house standards for photons of energies exceeding about 150 keV can be purchased, but they are not difficult to prepare from a radioactive solution (Section 4.2.2) and can then be expected to yield consistent countrates (allowing for decay). Source preparations are more difficult for β and for α particle sources (Section 1.3.5) when quite small chemical or physical changes in the source material could affect the extent of source self-absorption and so the reproducibility of countrates.

2.3 Radioactivity standards

2.3.1 Comments on their production and their purpose

Radioactivity standards are produced in national standard laboratories using procedures that are continuously reviewed and improved upon. A few of these procedures will be introduced in subsequent chapters, but on the whole the preparation of radioactivity standards is beyond the scope of laboratories concerned with radionuclide applications in industry and technology.

The large majority of radioactivity standards that are now sold are samples of solutions of radionuclides whose activity concentrations have been certified by a national or international standard laboratory or by authorised commercial laboratories. These standards serve to verify the activity of other sources by direct comparisons in suitable conditions and to measure the efficiency of detectors for particular radionuclides. They are also used to effect the calibration of SSIs.

Radioactivity standards are sold with a certificate listing the radioactivity concentration in becquerels per gram of solution due to the principal radionuclide at a stated reference date together with its uncertainty and confidence level (Section 6.5.1). The certificate should also give adequate information about the physical and chemical characteristics of the solution and about radionuclidic impurities.

Many industrial applications of radionuclides require accurate radioactivity measurements (often within $\pm 1.0\%$ or better) which cannot be done without help from standard sources or SSIs. Activities of radionuclides have to be known, both to realise satisfactory applications and also to realise adequate radiation protection.

2.3.2 The international dimension of radioactivity standards

International compatibility is essential for all types of physical measurements. It is for this reason that national standard laboratories around the world cooperate under the umbrella of the International Committee of Weights and Measures (CIPM, abbreviated from the French) which maintains a laboratory at Sèvres near Paris, France, the International Bureau of Weights and Measures (BIPM).

The scientific work of the CIPM is conducted by Consultative Committees. The Consultative Committee of interest here deals with standards for the measurement of ionising radiations (CCEMRI). Section I of the CCEMRI is concerned with establishing standards required for dosimetry measurements

(Section 2.4) and with verifying results of dose rate measurements obtained in standard laboratories. Section II of the CCEMRI looks after standards for measurements of radionuclide activities (Rytz, 1983).

The dosimetry standards used for such work are high-precision ionisation chambers and calorimeters set up in national standard laboratories and also at the BIPM. They serve for the calibration of SSIs e.g. those designed for use by radiation protection personnel. Dosimetry standards are also important for industrial applications, e.g. to ensure accurate results for chemical dosimetry (Section 7.6.2).

An important part of the work of CCEMRI Section II is to verify international compatibility of radioactivity measurements. To effect this, the BIPM organises international comparisons which serve to verify that the results of radioactivity measurements made on internationally distributed samples of solutions of a selected radionuclide are internally consistent. All participants are expected to return results for the radioactivity concentration of these samples which are the same within the stated uncertainties. Even though this aim is not always fully realised, the BIPM obtains evidence of how much compatibility can be expected in the prevailing circumstances.

2.4 Radiation dosimetry for radiation protection

2.4.1 Absorbed dose limitations

Radiation dosimetry is the science of the measurement of the effects of ionising radiations absorbed in any material including living organisms. The energies absorbed from these radiations are known as doses. Dosimetry serves, among others, to verify that doses of ionising radiations absorbed by experimenters or members of the public are well within specified limits and to warn that remedial action should be taken if these doses are exceeded. The following comments will deal with doses due to α, β and γ rays. Protection from doses due to neutrons will be discussed in Section 7.4. Mandatory precautionary provisions are not unique to work with ionising radiations. They apply to work with all substances that could be harmful to health if not used correctly.

Radioactive materials form a special category of hazardous substances. Health risks are easy to overlook because the emitted radiations are imperceptible to the unaided human senses, certainly at levels normally employed for industrial and similar applications. Yet relatively simple instruments are readily available (Section 2.6.5) which can detect ionising radiation increments of only a few per cent above the always present background radiations, thus permitting quick remedial action if called for. That such precautions are effective is demonstrated by the excellent health record of radiation workers (Luckey, 1998).

Radiation protection is organised on a world-wide basis and is supported by national governments and international organisations, e.g. the World Health Organisation (FAO/IAEA/WHO, 1981), the International Atomic Energy Agency (IAEA, see below), the International Commission on Radiation Units and Measurements (ICRU, 1993) and, above all, the International Commission on Radiological Protection (ICRP, see below). The ICRP has been and is the body responsible for most recommendations on radiation protection that are almost invariably adopted by national governments. Its recommendations are published as the *Annals of the ICRP*, for example ICRP (1991). Basic radiation safety standards are also published by the IAEA (IAEA 1986a,1986b,1996). They are regularly updated and are now available on the World Wide Web (Appendix 3, Table A3.1).

2.4.2 Units for exposure, absorbed and equivalent dose

The first internationally accepted radiation unit was based on the capacity of X or γ radiations to ionise air, a process known as exposure. The unit was named the röntgen (R) after the discoverer of X rays. It is the quantity of ionising radiation producing ions (of one sign) with a charge equal to 2.58×10^{-4} C/kg of NTP (normal temperature and pressure) air. An exposure of one röntgen is approximately equal to that of an absorbed dose of 0.01 gray (see Eq. (2.3)).

Since doses of radiations have biological as well as physical and chemical effects, it is necessary to distinguish between two types of measurements of absorbed dose.

(i)   Measurements of energy transferred from the radiation to unit mass of any absorbing material and for any purpose.

(ii) Measurements of energy transfers from the radiations to unit mass of living tissue, allowing not only for effects referred to in (i) but also for potential radiobiological and clinical consequences of the absorbed energies.

Prior to the introduction of SI units, the ICRP had accepted two units to account for energy transfers from ionising radiations to matter – the rad for case (i) and the rem for case (ii), rem being short for röntgen equivalent man. The rad was defined as an energy transfer of 100 erg (the pre-SI unit of energy) absorbed per gram of irradiated matter.

The rem measured equivalent doses, i.e. absorbed doses in rads multiplied by a dimensionless quality factor Q, which accounts for the relative biological effectiveness (RBE) of the radiations responsible for the dose. Hence:

$$\text{equivalent dose (rem)} = \text{absorbed dose (rad)} \times Q. \qquad (2.2)$$

It was the introduction of equivalent doses that made it possible to distinguish between processes (i) and (ii) above. This distinction has since been tightened (Section 2.4.3).

With the introduction of SI units during the mid-1970s, the rad and rem were replaced by respectively the gray (Gy) and the sievert (Sv) named after two scientists who had made valuable contributions to radiation dosimetry. However, the old units, including the röntgen, remain in widespread use making it necessary to include their definitions along with the SI units.

The SI unit for exposure, which took the place of the röntgen, is the X unit. An exposure is equal to 1 X unit when ionising radiation produces 1 coulomb (C) of charge of one sign in 1 kg of NTP air, hence 1 X unit = $(1/2.58 \times 10^{-4})$R = 3876 R.

The energy transferred by ionising radiations into matter equals 1 gray when 1 joule (J) of energy is imparted into 1 kilogram (kg) of matter:

$$1 \text{ Gy} = 1 \text{ J/kg}. \qquad (2.3a)$$

Since $1 \text{ J} = 10^7$ erg and $1 \text{ kg} = 1000$ g it follows from the definition of the rad (100 erg/g, see above) that 1 rad = 0.01 Gy and 1 rem = 0.01 Sv. Thus, applying Eq. (2.2) for SI units one obtains:

$$1 \text{ Sv} = 1 \text{ (J/kg)} \times Q. \qquad (2.3b)$$

2.4.3 *Weighting factors, w_R and w_T*

Following scientific advances in radiation protection and the introduction of SI units, the biological effects of radiations were re-examined, causing the ICRP to recommend the replacement of the Q factor (Eq. (2.2)) by two weighting factors: the radiation weighting factor w_R to account for the biological effectiveness of the absorbed radiations and the tissue weighting factor w_T to account for the differences between the radiation sensitivities of different parts of the body (Table 2.1).

For uniform whole body irradiations one has $\Sigma w_T = 1$, but for non-uniform irradiations one has to allow for the w_T values of the exposed parts of the body (Table 2.1(b)) and calculate the equivalent dose as defined in Table 2.2. Table 2.2 also includes definitions of frequently used quantities required for radiation protection (see also Appendix 1 and IAEA, (1996a)).

Table 2.1. *Radiation weighting factors, w_R, the Q values they replaced and tissue weighting factors, w_T.*[a]

(a) w_R

Radiation		w_R	Q
β, X, γ		1	1
α		20	20
Neutrons			
Thermal		5	2
0.1	MeV	10	7.5
0.5	MeV	20	11
0.1–2	MeV	20	–
2–20	MeV	·5	–

(b) w_T ($\Sigma w_T = 1$)

Tissue	Fraction of risk	Tissue	Fraction of risk
Gonads	0.20	Liver	0.05
Colon	0.12	Œsophagus	0.05
Lungs	0.12	Thyroid	0.05
Stomach	0.12	Skin	0.01
Bladder	0.05	Bone surface	0.01
Breast	0.05	Others	0.17

[a] The w_R and w_T values were recommended by the ICRP (1991) and are now internationally accepted.

When an absorbed dose refers to case (i), (Section 2.4.2) it is stated in grays as defined in Eq. (2.3a). If the dose is absorbed in part or the whole of an organism one has case (ii) and the absorbed dose has to be qualified by the appropriate weighting factor(s) and expressed in sieverts (see below).

One has to allow not only for what used to be the RBE, now w_R, but also for the radiation sensitivities w_T of all tissue and organs that are within the radiation field (the w_T values of greatest interest are listed in Table 2.1(b)). One then has:

$$\text{equivalent dose (Sv)} = \text{absorbed dose (Gy)} \times w_R \qquad (2.4a)$$
$$\text{effective dose (Sv)} = \text{absorbed dose (Gy)} \times w_R \times w_T$$
$$= \text{equivalent dose (Sv)} \times w_T. \qquad (2.4b)$$

Equations (2.4a) and (2.4b) replace Eq. (2.2).

On applying Eq. (2.4a) for what can be considered a sufficiently uniform whole body dose ($\Sigma w_T \approx 1$, Table 2.1(b)) of 1 mGy due respec-

Table 2.2. *Glossary of terms used in connection with radiation dosimetry and radiation protection*
(selected from definitions published by Standards Australia (1998); see also Appendix 1).

Absorbed dose The energy absorbed by matter from ionising radiation per unit mass of irradiated material. For radiation protection purposes this dose is averaged over a tissue or organ or the whole body. The SI unit of absorbed dose is the joule per kilogram, with the special name gray (Gy).

Committed effective dose The effective dose that will be accumulated during the 50 years following the time of intake of radioactive material into the body.

Deterministic effects Effects on a biological system in which the severity of the effect varies with the dose and for which there is usually a threshold.

Dose Where used without qualification, this term is taken to mean 'effective dose' or simply absorbed dose.

Effective dose The product of the equivalent dose (in a tissue or organ) and the tissue weighting factor (w_T), summed over all the tissues and organs of the body. The SI unit is the joule per kilogram, with the special name sievert (Sv).

Equivalent dose The product of the absorbed dose (averaged over a tissue or organ) and the radiation weighting factor (w_R) for the radiation that is of interest. The SI unit of equivalent dose is the joule per kilogram, with the special name sievert (Sv).

Gray (Gy) The special name for the unit of absorbed dose. 1 Gy = 1 J/kg.

Radiation worker (occupationally exposed person) A person who, in the course of employment, may be exposed to ionising radiation arising from direct involvement with sources of such radiation.

Radioactive material Any substance that consists of, or contains any radionuclide, provided that the specific activity of such material is greater than 70 Bq/g, or such other value defined in relevant legislation.

Radiological hazard The potential danger to health arising from exposure to ionising radiation; it may arise from external radiation or from radiation from radioactive materials within the body.

Radiotoxicity The toxicity attributable to ionising radiation emitted by a radionuclide (and its decay products) incorporated in the human body.

Sievert (Sv) The special name of the SI unit for both equivalent dose and effective dose. 1 Sv = 1 J/kg.

Stochastic effects Effects on a biological system in which the probability of an effect rather than its severity is regarded as a function of dose, without a dose threshold. Examples are carcinogenesis in exposed individuals and hereditary effects in the descendants of exposed individuals.

Radiation weighting factor (w_R) A non-dimensional factor used in radiation protection to weight the absorbed dose. The value of each radiation weighting factor is representative of the relative biological effectiveness of that radiation in inducing stochastic effects at low doses.

Tissue weighting factor (w_T) A non-dimensional factor used in radiation protection to weight the equivalent dose. It represents the relative contribution of each tissue or organ to the total detriment due to stochastic effects resulting from uniform irradiation of the whole body.

tively to γ radiations and thermal neutrons, one obtains, referring to Table 2.1(a):

- for 1 mGy due to γ radiations ($w_R = 1$): the equivalent dose = 1 mGy × 1 = 1 mSv.
- for 1 mGy due to thermal neutrons ($w_R = 5$): the equivalent dose = 1 mGy × 5 = 5 mSv.

The assumption of uniform doses due to each of these radiations can be accepted if these doses are no more than a few times the background dose. This will be illustrated by an example in Section 2.7.2. More complex cases are beyond the scope of this book (but see Appendix A2).

2.5 Dose limits

2.5.1 The linear hypothesis and the ALARA principle

This section aims to draw the reader's attention to policies designed to ensure effective radiation protection. For an adequate treatment of this large subject, readers are referred to the references listed in Appendix 2.

Effective radiation protection has to distinguish between two types of effects caused by radiation: (1) deterministic effects that occur within hours or days, but only above a defined threshold (see Table 2.4 later) and which are a function of the intensity of the absorbed radiation (2) stochastic effects where radiation intensities determine probabilities. The more intense the radiation, the greater the probability that there will be harmful consequences to health, not just within hours or days as applies to deterministic effects but 12 to 24 months or even many years into the future. Stochastic effects include cancers and it is the avoidance of carcinogenic effects which is a principal concern of radiation protection.

After careful investigations the ICRP accepted the hypothesis of a linear, no-threshold relation between dose and stochastic effects right down to zero dose. For example, the probability of developing cancer due to the absorption of ionising radiations is not zero until the dose is zero, namely never. This hypothesis was accepted by the ICRP nearly three decades ago as the most conservative assumption in the circumstances and remains the accepted policy to this day (but see Graham *et al.*, 1999).

Given a no-threshold effect for the generation of cancers, it was clearly essential to reduce maximum allowed doses as much as practical. To achieve this objective without unduly restricting constructive applications, the ICRP recommended (back in 1973) that regardless of an allowed maximum, doses to operators and others should always be kept **As Low As Reasonably**

Table 2.3. *Dose limits for radiation workers and members of the public.*[a]

Type of dose	Limits[b]	
	Occupational	Public
Effective dose averaged over the whole body	20 mSv/year averaged over 5 years	1 mSv/year
Annual equivalent dose to:		
hands and feet	500 mSv	–
skin	500 mSv	50 mSv
lens of the eye	150 mSv	15 mSv

[a] The stated limits are maximum values. They must not be treated as allowed maxima but according to the ALARA principle (see Section 2.5.1). They could also be subject to qualifications issued by the local Department of Health.
[b] These limits apply over and above the background at the location of interest (Section 2.5.3), while doses due to medical treatment are not subject to statutory limitations.

Achievable (ALARA), taking account of benefits due to a procedure as well as any risks. This policy is known as the ALARA principle, but there is more to this policy than can be detailed here.

The linear hypothesis and the ALARA principle were adopted by all national radiation protection authorities and played their part in the rapid acceptance, by the same authorities, of all recommendations incorporated into ICRP Publication 60 (ICRP, 1991) which included a drastic reduction in allowed doses (see Table 2.3 for the currently allowed maxima). The reductions were accepted without significant effects on the steadily growing number of applications.

Typical examples demonstrating dose reductions while maintaining the level of applications are shown in Figures 2.2 and 2.3. Figure 2.3 shows results of radiation monitoring for four occupational groups obtained during 1991 in Australia which would have been typical for dose rates received then and since in other countries. By 1991 over 90% of Australian radiation workers registered doses which were less than 10% of the 20 mSv per year maximum (Table 2.3).

2.5.2 Deterministic and stochastic effects

As stated in Section 2.5.1, deterministic effects occur within hours or days if the absorbed dose exceeds a threshold (Table 2.4). Stochastic effecxts could become apparent months or even decades after exposure and are measured as probabilities – the higher the dose the more probable the effect. However, the

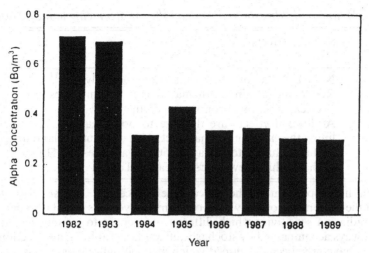

Figure 2.2. Concentration of α particles in the air of an Australian uranium mine. Currently achieved concentrations in Australian mines are over ten times less than the allowed maximum (UIC, 1998).

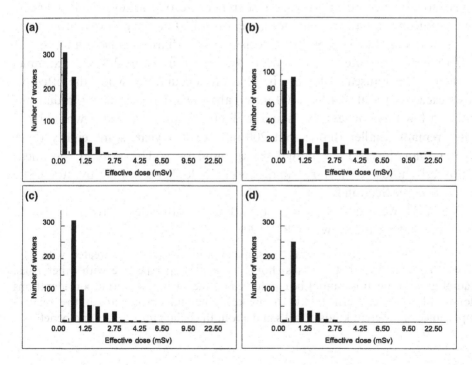

Figure 2.3. Effective doses absorbed during 1991 by Australian radiation workers from four occupational groups: (a) nuclear medicine workers; (b) industrial radiographers; (c) uranium miners; (d) mineral sands miners. The dose rate scale was chosen to permit a clear presentation of the data (Morris, 1992).

Table 2.4. *Whole body threshold doses capable of producing somatic effects.*[a]

Threshold dose (Sv)	Somatic effects
< 0.25	No detectable somatic effects, others unlikely
0.25 to 1	Reddening of skin (erythema), nausea in some cases
1 to 2	Nausea, vomiting, pronounced erythema
2 to 3	Additional effects are damage to bone marrow and to the cells lining the gastro-intestinal tract. Also neuromuscular damage
3 to 4	Lethal within 30 to 60 days in 50% of all cases (LD50)/60[b]
5 or more	Lethal within a few hours in nearly all cases

[a] Effects will not appear unless the stated dose is exceeded. The severity of effects depends on the dose rate, normally days or hours or even less and it also depends on individual radiation sensitivity which differs from person to person. Deterministic effects are always accompanied by stochastic effects (see Table 2.2 and Section 2.5.2).
[b] Treatments have now been developed which make lethal outcomes less likely even after very severe somatic effects.

probability of cancers following doses of the order of the average background is very low or, according to some researchers, zero (Luckey, 1998). Cancers due to ionising radiations may not appear until 20 or even 30 years following exposures, so greatly complicating decisions on radiation-caused cancers.

Dose limits for deterministic effects are shown in Table 2.4. For example, erythema and nausea following irradiations occur only when the effective dose exceeds about 0.4 Sv (400 mSv) absorbed from ionising radiations within a few days or less. No somatic ill effects are observed if whole body doses remain smaller than about 200 mSv over a year, some eighty times greater than the average annual background dose. However, if the linear hypothesis holds, even very small doses could be followed by stochastic effects (but see Section 2.5.3).

The ICRP went out of its way to underline both sides of the risk–benefit situation, stating as follows (ICRP, 1991, p. 67):

The Commission emphasises that ionising radiations need to be treated with care rather than fear and that its risks should be kept in perspective with other risks. Radiological protection cannot be conducted on the basis of scientific considerations alone. All those concerned have to make value judgements about the relative importance of different kinds of risks and about the balancing of risks and benefits.

2.5.3 *Background doses and their relevance for radiation protection*

Background radiations, estimated as averages for industrial countries, include both the natural and man-made background. They are listed in Table 2.5,

Table 2.5. *Average dose rates due to the natural and man-made background of ionising radiations in industrialised countries.*[a]

Type of radiation	Estimated dose rate (mSv/year)
Dose rates from naturally occurring radioelements in the soil, rocks, etc.	0.40
Dose rates due to radon, thoron and their short-lived decay products	1.20
Cosmic radiation at sea level	0.25[b]
Radionuclides inside the human body, mainly potassium-40	0.30
Doses due to medical procedures, mainly X rays	0.30[c]
Doses due to other man-made sources (consumer products, air travel, fall-out, etc.)	0.05
Total, at sea level	2.5

[a] For comments on the entries in this list see Section 2.5.3.

[b] Dose rates due to cosmic rays are functions of latitude as well as height above sea level. At 1000 m and at temperate latitudes the dose is twice, and at 4000 m it is about seven times the stated values.

[c] The preceding entries refer to radiations of natural origin. This and the next entry are due to man-made sources.

adding up to 2.5 mSv per year. This result is based on figures published by the UK National Radiological Protection Board (NRPB) and UNSCEAR (1988).

Doses due to ionising radiations in any region cannot be lower than the dose due to the background (though it may be practical to reduce the man-made component of the background). Background radiations depend, amongst others, on height above sea level and so vary from place to place by upwards to factors of three and more. Yet there is no conclusive evidence of corresponding changes in deleterious effects to health.

In parts of India and China and in parts of Cornwall in the south-west of England (Clarke, 1991, Graham *et al.*, 1999, McEwan, 1999), populations numbering several millions live subject to background levels in the range 3 to 8 mSv per year and higher for smaller groups. These relatively high dose levels are due mainly to radiations from geological structures rich in uranium, thorium and their decay products (Figure 1.5).

Similarly high doses are experienced in cold climates, resulting principally from the accumulation of radon and its short-lived decay products in homes sealed against the cold. Numerous investigations have been made and are continuing to identify differences in carcinogenic effects that could be assigned to these relatively large differences in the ionising radiation background but so far without conclusive results.

The differences in background radiation intensities are large in relative terms but evidently small as regards consequences to human health. To effect a reliable identification of possible differences in cancer rates, i.e. a higher rate in higher background regions, would require far more sensitive, statistical tests than have been practical so far. In part this is so because ionising radiations are a minor cause of fatal cancers. According to data published by the UK NRPB (Clarke, 1991), diet causes 35% of all fatal cancers in the UK, tobacco smoking causes 30%, but ionising radiations cause only 1%. These figures would not be greatly different in other industrialised countries.

Current estimates of health risks due to the natural background, as discussed so far, are based on the validity of the linear, no-threshold hypothesis (Section 2.5.1). However, since background radiations have been present as long as life on earth, background effects have also been studied from the viewpoint of natural evolution. This latter approach has so far remained controversial with the ICRP adhering strictly to established policies. Nevertheless, the evolutionary approach should be mentioned, though this can only be done in very brief outline.

Natural background radiations, more intense than those causing the doses recorded in Table 2.5, were present when the earth was formed, slowly decreasing in intensity to their present level. It has been proposed (see e.g. Parsons, 1992, 1999), that natural evolution has adapted all living organisms to the prevailing level of background radiation just as it did to other normally prevailing factors, e.g. climate or food. The evolutionary adaptation of organisms to stresses to which free living populations are exposed is often known as hormesis.

In a recent report prepared for the International Nuclear Societies Council, Graham *et al.* (1999) put forward strong evidence in favour of the possibility and significance of bio-positive effects from low-dose radiation exposures (<20 mSv per year) which, they propose, need to be accepted without prejudice (see also Keay, 1999).

However, until the ICRP recommends changes in its current radiation protection policies and these changes are generally accepted, it remains essential to abide strictly by all current ICRP recommendations and in particular the linear, no-threshold hypothesis and the ALARA principle; this is done throughout this book.

2.6 Radiation protection in the laboratory

2.6.1 Classifications of sources and of laboratories

There is a large body of literature on safety in laboratories, especially in laboratories used for work with ionising radiation. This literature is made available by several international organisations, notably those mentioned in Section 2.4.1. In Australia, safety rules are published as standards drawn up by committees of experts brought together by Standards Australia, the Australian national standards organisation (Standards Australia, 1998). These standards cover all aspects of the subject, are concisely written and periodically updated. A similar situation would apply in other countries.

Radioactive materials could be of low enough activity for their radio-activity to be disregarded. This applies, as a rule, to radionuclides of specific activity below 70 Bq/g, or somewhat higher for a few non-toxic radionuclides. For details about sources that can be treated as non-radioactive, experimenters are referred to their national radiation protection authority. When radionuclides are employed in a laboratory risks to the health of workers are principally functions of the intensity and the energy of the radiations but there are also other characteristics of radioactive materials which have to be considered, notably their chemical nature. Radionuclides that are given a low hazard rating have a simple chemistry, emit radiations with energies within a few hundred kiloelectronvolts and half lives of no more than two to three months.

National and international authorities have classified radionuclides into four groups depending on their hazard level: very high, high, moderate and low. Laboratories and other workspaces are graded by their facilities permitting sufficiently safe handling of high or not so high radioactivities and, in particular, unsealed radioactivities (Mann *et al.*, 1991, Table 3.3). They are then labelled high-level, medium-level or low-level laboratories for each of the four groups, e.g. high activities of radionuclides classified as creating very high hazards (e.g. high-activity liquid sources), may only be used in a laboratory rated for high-level sources. In addition it is always essential to ensure that the amount of radioactivity is minimised and that radioactive materials are safely transported and stored (IAEA, 1996b).

Facilities in low-level laboratories are not greatly different from those in laboratories that do not employ radionuclides. However, the former have to be equipped with a fume cupboard (with isolated exhaust), a shielded store for radionuclides and also other more specialised shielding (Section 2.6.2) and more specialised equipment, notably a secondary standard instrument

(Section 2.2.3) and radiation monitors (Section 2.6.5). Other rules refer to common sense matters such as absolute cleanliness, use of protective clothing, banning of food, drink, smoking and cosmetics, clear marking of the activity of all radioactive substances and other record keeping as well as the handling and disposal of radioactive waste (Section 2.6.3).

Practioners should verify with their local authorities about the adequacy of their laboratory facilities for the types of radionuclides and activities likely to be employed there.

2.6.2 Time, distance and shielding

The three most effective rules consistently employed by radiation workers to minimise exposure when using ionising radiations are: minimise the time close to radioactive sources, maximise the distance away from them and make full use of shielding. The effectiveness of these rules and related measures is apparent from the data in Figure 2.3 (see also Section 2.6.6).

Shielding adds to the effectiveness of both the time and distance rule and is mandatory in many situations. Radioactive sources should be shielded in lead pots of appropriate thickness, allowing for the energies and the intensities of the emitted radiations (see below). Stores of radionuclides should be shielded in firmly locked steel cupboards placed out of the way, thus adding distance shielding to metal shielding. In every case shields should be thick enough to ensure that absorbed doses at locations accessible to the general public are within prescribed limits.

Whenever practical use should be made of commercially available bench top shields, 2 to 5 cm thick lead, fitted with lead glass visors as well as of workstations for use with β particle emitters. They are fabricated from 10 mm thick acrylic sheets, 50 cm high on a plastic base, containing stands for test tubes, pipettes, long-handled tweezers, etc. with very smooth, hard surfaces to facilitate the removal of contaminants. Manufacturers of this equipment also offer other devices to improve the effectiveness of shielding in situations commonly met during industrial applications. A list of suppliers of nucleonic equipment is given in Section 4.3.3.

The effectiveness of shielding introduced to attenuate γ radiation is commonly measured in terms of half value layers (HVLs). This is the thickness of a material which reduces the intensity of a beam of γ rays to half its initial intensity, leading to a corresponding reduction in dose. HVLs for a few commonly used materials can be found in Table 7.3. HVLs for γ ray energies and Z numbers intermediate between those in the table can be estimated by linear interpolation.

2.6.3 Coping with radioactive waste

Waste is, by definition, material for which no further use is forseen. There are three rules for coping with unwanted radioactive materials. They are summarised as:

concentrate and contain
delay and decay
dilute and disperse.

Radioactive materials which are no longer useable but have a long half life (say longer than three months) and are still significantly active should be reduced in volume (concentrate) and stored behind sufficiently thick shielding preferably in an out of the way location (contain). If the half lives of waste materials are two to three months or less it is usually practical to keep them in shielded storage (delay) until the activity has decayed enough to permit disposal as inactive waste. The 'dilute and disperse' procedure refers to relatively low activity liquid waste in an aqueous solvent which can be sufficiently diluted, using the same solvent, to be dispersed into the sewer. It is then essential to realise a radioactivity concentration and a total activity which are low enough to be accepted by the local sewerage authority. Provided this can be done, the diluted radioactive liquid can be safely dispersed into the sewer, if not all at once then in one or more volumes separated by prescribed intervals.

An option which should be preferred when unwanted sources are expected to remain comparatively or highly active for a year or longer is to return the material to the supplier or to a facility for active waste storage such as a nuclear power station or a nuclear research laboratory.

The disposal of radioactive waste is a sensitive matter, and care must be taken to comply with all regulations and guidelines (see below).

2.6.4 The radiation advisory officer

If an organisation employs radionuclides on a more extensive scale, it is customary and mandatory in some countries to appoint a person with specialist knowledge in health physics and related topics to act as radiation advisory officer (RAO). This requires attendance to what are often complex problems relating to all aspects of radiation protection. The duties of RAOs are listed in all publications on radiation protection (see Appendix 2).

The RAO has to take control during emergencies such as spillages of radioactive material, accidental ingestions, fire or other accidents. On a less critical level, the RAO has to keep track of all licensable radioactive materials from the time they have been received to the time they are disposed of as radioactive waste or otherwise.

The RAO is, as a rule, responsible for the maintenance and calibration of all radiation monitors (Section 2.6.5) and instructs other staff on their use. This includes arrangements for the correct employment of personal dosimeters. The RAO takes the necessary steps to ensure that radiation levels in laboratories and adjacent rooms are kept within prescribed limits and that there are adequate facilities for personnel to work safely and for other staff and visitors not to be exposed to higher radiation levels than permissible. The RAO also attends to the management of radioactive wastes and is the contact officer for communications with the government authority appointed to supervise radiation protection regulations.

2.6.5 Radiation monitors

Radiation monitors, calibrated at regular intervals against appropriate standards and chosen to cope with all types of measurements likely to be required, have to be kept by law in any laboratory or workplace where ionising radiations are employed in any form. They have to be available in addition to personal dosimeters which have to be worn by all workers likely to exceed an annual maximum dose as stated by the local authorities (Table 2.3).

The long preferred type of personal dosimeter is the thermoluminescent dosimeter (TLD). It contains a thin strip of lithium fluoride (LiF) or material with similar properties whose atomic electrons are excited by the radiation and remain in an excited state until de-excited by subjecting the LiF to a prescribed heating cycle. The processing is done periodically in a laboratory authorised to carry out these measurements. The TLD badges are designed so that contributions from the background and from other sources of radiation accumulated while the monitor was not worn can be excluded from the total.

With TLD readings usually available only at four weekly or longer intervals, one relies on radiation monitors to verify that dose levels are within the limits prescribed for the location, that all surfaces (including the skin and clothing of personnel) are free from contamination and that no radioactive material was accidentally absorbed into any body.

Numerous types of portable monitors are now commercially available, many of them miniaturised and employing micro-processor circuits, digital read-outs, automated background subtraction and alarms. It has to be borne in mind that monitors register countrates at the point where they are used. If one requires the decay rate in a source, the monitor has to be calibrated with a standard of the radionuclide of interest and used in the appropriate source-detector geometry (Sections 6.3 and 6.4).

Commonly used portable monitors are Geiger–Müller (GM) counters, proportional counters, scintillation detectors which are, as a rule, NaI (Tl) crystals, ionisation chambers and solid state detectors. The comments that follow deal with a few specialised aspects of instruments used for radiation protection. A discussion of the principles by which they work is given in Sections 5.2 and 5.3.

Depending in part on the size and thickness of their windows and other aspects of their geometry, GM and proportional counters can be almost 100% efficient when detecting charged particles. However, since the detection medium is in each case a gas at or below atmospheric pressure, the γ ray efficiency of these detectors is only about 1% though it increases significantly when detecting photon energies below about 70 keV. The low efficiency for background radiations, which are nearly all γ rays, explains why these detectors can often be used without shielding (Section 4.6.4).

Most GM detectors are small tubes (Figure 5.3(a)). Problems with their dead time, which is rarely less than 100 μs, can be encountered. Proportional counters are the more efficient surface monitors but they are also more expensive to purchase and use. Their sensitive area can be 200 cm² or more and their dead time is only a few microseconds. Another specialist type of portable proportional counter serves for the detection of neutrons (Section 5.4.4) and there are also designs for the efficient detection of low-intensity α particles (Section 4.3.3).

NaI(Tl) crystals are close to 100% efficient for the detection of low-energy γ rays (about 20 to 100 keV) though less so with increasing γ ray energies (Figure 4.4). Shielding these crystals from background radiation is discussed in Section 4.6.4. Furthermore, α, β and γ radiations can now be detected using semiconducting detectors (Section 5.5.2). These detectors have also become available as personal dosimeters. This is an important fact since this latter type of dosimeter can be read continuously by the user, so providing direct, up-to-date information on absorbed doses rather than on a monthly basis thanks to the services of a specialist laboratory.

Portable ionisation chambers, are also available and, being free of dead time, are particularly useful for the monitoring of high activities of ionising radiations. All these instruments and how they work are described in detail in the catalogues of manufacturers several of whom are listed in Section 4.3.3.

2.6.6 Guarding against radioactive contamination

Radioactive sources used for applications should be sufficiently safely encapsulated to avoid contamination of laboratory space or operators (Section 4.2.3). However, this is rarely practical when having to dispense radioactive liquids (IAEA, 1996a). When this is the case, the dispensed activities should be kept as small as practical, preferably below the mega-becquerel level, but the use of higher activities is sometimes unavoidable, e.g. for nuclear medicine applications or when adding radioactive tracer solution to a large volume of liquid (see e.g. Section 8.4.4).

There is, of course, the general requirement for maximum cleanliness, strict and clear records, protective clothing and the other precautions referred to in Section 2.6.1.

Work with radioactive liquids almost invariably leads to problems arising from spillages. It should always be done on a strong plastic tray, similar to a large photographic development tray, lined with absorbing material to contain accidental spillage. All the steps of such operations should be carefully prepared, as should all work with any unsealed radioactive materials (Section 2.6.2). Precautions are particularly important if there is any risk of radioactive material entering the body of an operator; careful tests are called for if this should happen.

Vessels prepared for dilutions should be placed so that they cannot be upturned and sealed vessels should be centrifuged before being opened to ensure that no liquid is left near the neck. Once opened, they should be stored inside another vessel. Syringes used for radioactive liquids should not be used more than once to avoid cross contamination. All exposed surfaces, not least the skin and clothing of experimenters, should be periodically monitored to detect contamination and the tray should be large and strong enough to accommodate shielding material should that be called for. There is here another advantage of working with short half life radionuclides – contamination of equipment is soon eliminated.

2.7 Dose rates from alpha, beta and gamma ray emitting radionuclides

2.7.1 Rules-of-thumb for work with alpha and beta particles

This section will briefly list a few approximations and rules-of-thumb applicable when working with α or β particles. Section 2.7.2 covers approximations applicable when working with X and γ rays. A list of rules-of-thumb was reported by Shleien (1992, Ch. 3).

Precautions based on approximations can be accepted as safe when working with ionising radiations provided one can be confident that doses absorbed annually by operators are not much greater than 20–30% of the currently allowed annual maximum dose for radiation workers (Table 2.3). The results shown in Figure 2.3 show that this condition can be realised for the large majority of radiation workers.

Members of the public would not normally be allowed to come near enough to radiation sources to be affected by doses that could cause detriment to health, except during medical procedures which are currently treated in a category of their own – they are not subject to ICRP limitations. What is acceptable in any given case is judged by experienced nuclear medicine physicians.

Doses due to α particles can be estimated using the approximation that these particles lose about 1 MeV/cm travel in air (Figure 3.2(a)) or proportionately more in denser materials. The kinetic energy of α particles has to exceed about 7.5 MeV for them to penetrate the dead layer of the skin (0.07 mm on average).

To estimate energy losses of β particles and other electrons, the following rules-of-thumb apply per 10 mm travel:

in aluminium ($\rho = 2.70$ g/cm^3) 5.0 MeV
in lucite ($\rho = 1.2$ g/cm^3) 2.3 MeV
in tissue or water ($\rho = 1.0$ g/cm^3) 2.0 MeV
in room air ($\rho = 0.0013$ g/cm^3) 0.0025 MeV.

Energy losses in other materials can be estimated as proportional to their physical density. Beta particles emitted from radioactive sources about 0.3 m distant from experimenters rarely require more than about 3 mm of plastic material between the source and the skin to prevent penetration.

If β particles penetrate into tissue, they impart a relatively high dose, approximately 0.3 mSv per 1000 betas per hour on average. This rate is only weakly dependent on β particle energy. It demonstrates the need to handle β particle sources using plastic gloves and long-handled tweezers. Electrons should be absorbed in plastic material (to minimise bremsstrahlung radiation, Section 3.8).

2.7.2 Dose rates from X and gamma radiations

The dose equivalent rate constant, D_{eq}

Doses due to γ rays are normally measured in terms of the dose equivalent rate constant D_{eq}, recalling that the term equivalent dose refers to doses absorbed into living organisms (Eq. (2.4a)).

Measurements of doses due to absorbed X or γ rays are difficult because of physiological and anatomical complexities applying to irradiated individuals. These difficulties are reduced by making absorbed dose measurements in air at the surface of the skin and allowing for the difference in the mass energy absorption coefficients of air and tissue to calculate doses to unit mass of tissue.

Calculations of doses to unit mass of air employ the air kerma rate constant for the radionuclide of interest, say R, here written $(D_{KE})_R$. These constants are measured in standard laboratories as functions of the decay scheme of each radionuclide and of the density and other properties of air. The term kerma stands for the **k**inetic **e**nergy of charged **r**adiations (commonly electrons dislodged by X or γ rays) released per unit **m**ass of **a**ir.

To calculate doses to tissue from doses to air, one employs the dose equivalent rate constant D_{eq} referred to earlier. For routine purposes using low doses, D_{eq} can be taken as equal to 1.1 D_{KE}, ignoring variations from one individual to another but assuming that the γ rays pass right through the body unless there are interactions with tissue. D_{eq} is expressed in the same units as D_{KE} except that a dose to tissue is measured in sieverts not in grays (Section 2.4.2). Table 2.6 lists values of D_{eq} for a number of widely used radionuclides selected from a list published by Wasserman and Groenewald (1988, see also Groenewald and Wasserman, 1990). Other references are listed in Appendix 2.

Dose calculations

With D_{eq} in the units used in Table 2.6 and assuming a whole body dose due to a quasi-point source d metres from the experimenter, a scatter-free environment and ignoring the background dose, the whole body dose to experimenters, here written D_g, due to, say, a cobalt-60 source of activity A MBq can be calculated as:

$$D_g = [(D_{eq})_{Co60} \times (A/d^2)] \, \mu Sv/h. \tag{2.5}$$

Let us assume that an experimenter employs a 5 MBq point source of ^{60}Co to effect a large number of attenuation measurements employing the precautions outlined in Section 2.6.2. Let the preparatory arrangements require 2 hours per month, when the experimenter has to be at 0.5 m from the unshielded point source (each operation takes no more than a minute or two, but the total per month is as stated), and let these measurements be continued for a year. Using D_{eq} for ^{60}Co from Table 2.6 and entering the data into Eq. (2.5) yields:

Table 2.6. *Dose equivalent rate constants (D_{eq}) for soft tissue irradiated by selected radionuclides.*[a]

Radionuclide	$m^2 \mu Sv/(MBq\ h)$
Carbon-11	0.154
Fluorine-18	0.149
Sodium-24	0.473
Potassium-42	0.035
Chromium-51	0.0043
Iron-59	0.161
Cobalt-57	0.015
Cobalt-60	0.337
Zinc-65	0.073
Technetium-99m	0.016
Iodine-131	0.056
Caesium-137	0.086
Tantalum-182	0.177
Iridium-192	0.121
Gold-198	0.062
Mercury-203	0.035

[a] Based on results quoted by Wasserman and Groenewald (1988). The figures apply for photon energies above 20 keV. Uncertainties are in the range ± 2 to $\pm 6\%$, small enough for radiation protection purposes.

$$D_g = 0.337 \times 12 \times 2 \times 5/(0.5)^2 = 162\ \mu Sv \approx 0.16\ mSv.$$

The approximately 10% per year decay of ^{60}Co has been ignored. Taking the annual background dose as 2.5 mSv (Table 2.5), the ^{60}Co dose absorbed by the experimenter was only about 6% of the background. Even if greatly underestimated the dose would still have less effect on experimenters than many changes in location.

When working with a radionuclide of known decay scheme but unknown D_{eq} value, whole body doses to operators (D_g) could be obtained from the rule-of thumb expression:

$$D_g = [0.14 \times \Sigma(f_\gamma\ E_\gamma) \times (A/d^2)]\ \mu Sv/h \qquad (2.6)$$

(Shapiro, 1972, Ch. 13), subject to the same simplifying assumptions as applied to Eq. (2.5). On comparing equations (2.5) and (2.6), one obtains:

$$D_{eq} = 0.14 \times \Sigma(f_\gamma E_\gamma)\ \mu Sv/(MBq\ h), \qquad (2.7)$$

where f_γ is the fraction of γ rays of average energy E_γ MeV.

For ^{137}Cs decays one has $f_\gamma = 0.85$ and $E_\gamma = 0.66$ MeV (Figure 3.4(d)). Ignoring the X ray contributions as relatively insignificant, one obtains

$D_{eq} = 0.14 \times 0.8 \times 0.66 = 0.079 \text{ m}^2 \text{ } \mu\text{Sv/(MBq h)}$, which agrees with the value of D_{eq} for ^{137}Cs in Table 2.6 within 10%. However, when doses to experimenters are likely to exceed the allowed maximum (20 mSv per year), it would be advisable to ensure that absorbed doses are calculated with as much accuracy as necessary.

Chapter 3

Properties of radiations emitted from radionuclides

3.1 Tools for applications

This chapter will introduce radionuclides and the emitted radiations as tools for applications. We shall begin with the well known α, β and γ radiations followed by a few characteristics of extranuclear electrons and X rays. Additional comments about neutrons (see Section 1.3.6), will follow in Section 5.4.4.

3.2 Properties of alpha particles

3.2.1 The nature and origin of alpha particles

Alpha particles are here called primary radiations because their emission is the first evidence of nuclear disintegrations which turn parent atoms into daughter atoms belonging to a different chemical element. Alpha particles are emitted either as a single branch, i.e. all with the same energy, or as several branches each with its own energy. A typical case is the α particle decay of americium-241. Close to 13% of the decays occur at an α particle energy of 5.44 MeV and close to 85% at 5.48 MeV, leaving three minor branches each with its own energy, overall exactly one α particle emitted per decay.

Properties of α particles were summarised in Table 1.2. With $Z = 2$ and $A = 4$, α particles are physically identical to the nuclei of helium atoms. As calculated by Rutherford (Section 1.2) the diameter of α particles is about 10^{-14} m, or some 10^4 times smaller than the 10^{-10} m atomic diameters. But, given a mass equal to that of four nucleons, ($m_\alpha \approx 3730$ MeV as against 0.51 MeV for m_β), they are only emitted from high Z number radionuclides when they reduce the mass of the parent by four units and its Z number by two units.

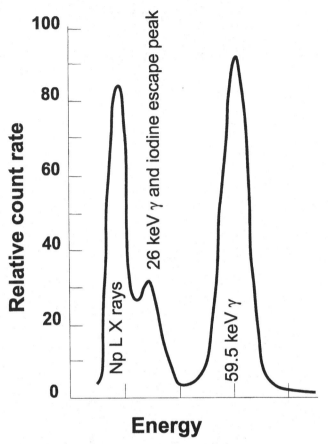

Figure 3.1.   The photon spectrum following α particle decays of americium-241 to neptunium-237, obtained with a NaI(Tl) detector fitted with a thin Be window. The spectrum is commonly assigned to americium-241 but is due to X and γ rays emitted from its daughter neptunium-237. The ^{237}Np LX ray peak near 17 keV is the average of four LX rays. The 26.3 keV γ ray peak is unresolved from the about 30 keV iodine escape peak, but the 59.5 keV γ ray peak is clearly resolved (Section 6.4.2).

Following their emission with high energies, mostly between 3 and 8 MeV, α particles forge through the negatively charged electrons of neighbouring atoms in straight lines, exerting Coulomb forces as they dissipate their kinetic energy by exciting electrons and ionising atoms. Deflections from straight lines occur only near the end of their paths.

The α particle emitter americium-241 (^{241}Am) is frequently selected for applications in science and industry because its α particles are followed by γ rays with the often useful energy of 59.5 keV (Figure 3.1) though only at a rate of about 36 γ rays per 100 decays. Other useful characteristics are a relatively simple chemistry and the fact that its daughter neptunium-237

(^{237}Np) is effectively stable ($T_{1/2} = 2.1 \times 10^6$ y) as compared with the daughters of one of the numerous α particle emitters which belong to the family of naturally radioactive elements (Figure 1.5). Notably, there is radium-226, with its short-lived daughters, and the polonium isotopes. They have to be used with care: one must distinguish between the α particles emitted from the short-lived daughters of radium-226 while polonium isotopes have difficult chemistries.

3.2.2 Alpha particle interactions with matter

Alpha particles and other nuclear radiations (neutrinos are an exception, see Table 1.2) ionise the atoms of the material through which they travel either directly or indirectly (Section 1.3.5), being known as 'ionising radiations' for this reason. Following an ionisation, ions of opposite charge rapidly recombine unless they are forced apart and collected by an electric field.

Alpha particles and other ionising radiations expend only part of their energy on ionisations. The balance is spent on exciting outer atomic electrons to higher energy levels which promptly de-excited when the energy is emitted mostly as photons of light and eventually as thermal energy. The average energy expended by α particles (and also by β particles) to create ion pairs in a gas is between 20 and 40 eV per ion pair, depending mainly on the nature of the gas, with some 50 to 60% of this energy going to excitations.

Alpha particle energies are quickly dissipated because, being doubly charged, these particles interact intensely with the vast numbers of atomic electrons through which they pass (Section 3.2.3), so accounting for their short but well defined ranges (Figures 3.2(a), 3.3(a)) at the ends of which they each capture two electrons to continue as atoms of helium gas. As mentioned in Section 3.2.1, α particles only deviate from straight lines near the end of their paths (Figure 3.3(a)). Alpha particle sources should be made as thin as possible to minimise energy loss in the source, which happens all too readily prior to emission (Section 4.2.2). It is also useful to note that the half lives ($T_{1/2}$) of radionuclides decaying 100% by α particle emissions are likely to be proportional to their mass (m). Taking the case of equal activities of ^{226}Ra and ^{241}Am, one has:

$$m_{Am}/m_{Ra} = (T_{1/2})_{Am}/(T_{1/2})_{Ra} = 433 \, \text{y}/1601 \, \text{y} \approx 0.27. \tag{3.1}$$

Assuming similar chemistry, sources of ^{241}Am are likely to be thinner and so subject to less self-absorption than sources of equal activity of ^{226}Ra.

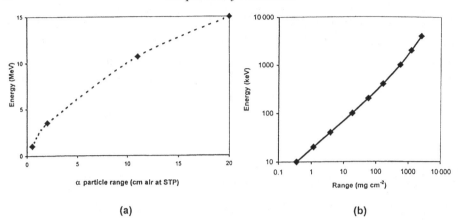

Figure 3.2. (a) Ranges of α particles (in centimetres) in room air at standard pressure (760 Torr) as a function of particle energy in MeV. (b) Ranges (approximate) of electrons (in mg/cm^2) in light materials. The ranges for β particles refer to the maximum energy of the spectrum.

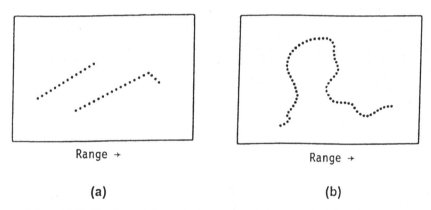

Figure 3.3. (a) Following their emission, α particles travel through a photographic emulsion in straight lines but could be scattered near the end of their range. (b) Beta particles move along tortuous paths but they have well defined ranges as shown in Figure 3.2(b).

3.2.3 Ionisation intensities of alpha particles

The average energy needed by α and also by β particles to create an ion pair in air is 35 eV. The range of 6 MeV α particles in air at standard temperature and pressure (STP) is just over 50 mm (Figure 3.2(a)). Over that range an α particle creates, on average, about $6 \times 10^6 / 35 = 1.7 \times 10^5$ ion pairs and therefore about 3400 ion pairs per millimetre path, a very high ionisation density. The ionisation intensity due to β particles is about 10^3 times lower than that, but their range is of order 10^3 times longer.

The range of α particles in materials is approximately in inverse proportion to their density. In tissue, the range is, on average, less than 0.02 mm/MeV, yet α particles each create some 10^6 ion pairs over this short distance, so explaining their high relative biological effectiveness and high radiation weighting factor (Table 2.1(a)).

The rates of ion pair formation quoted above are only approximate. Ionisation intensities vary with α particle energy and also for other reasons. If accurate values are needed they have to be obtained from health physics texts (see Appendix 2).

3.3 Properties of beta particles

3.3.1 Beta particles and electrons

Beta particles are electrons emitted from the nuclei of radioactive atoms. They are known as primary radiations for the same reason as α particles. Like α particles, β particles are emitted either as a single branch (Figure 3.4(a), (b)) or as fractional branches. Figure 3.4(c) shows the two-branch-decay of caesium-137, 95% to an excited state of the daughter ^{137m}Ba and 5% to its groundstate ^{137}Ba. The de-excitation of ^{137m}Ba is yet to be introduced.

Beta particles are physically identical to atomic electrons. However, atomic electrons are negatively charged whereas β particles could be either negatively or positively charged when they are known as negatrons or positrons respectively (Table 1.2).

Decays by negatron emissions give rise to daughter nuclides with a Z number one unit higher than that of the parent. Decays with the emission of positrons or following the capture by a nucleus of one of its atomic electrons (electron capture (EC), Section 3.6.1), lead to daughters with a one unit lower Z number than the parent. Many radionuclides decay by both of these processes and a few by all three decay modes as will be seen in Section 3.7.4.

Beta particles, both negatrons and positrons, are emitted with a spectrum of energies the shape of which is a property of the emitting radionuclide (Section 3.3.4). Negatrons are stable elementary particles. In contrast, positrons exist for only about a picosecond (10^{-12} s). Like negatrons they quickly lose kinetic energy during scattering events with atomic electrons. On approaching thermal equilibrium, they acquire a rapidly increasing probability for an annihilation reaction with omnipresent negatrons. Invariably annihilations occur, when $e^+ + e^- \rightarrow 2\gamma$, known as annihilation γ rays which carry away the rest-mass energy of the two particles, 511 keV each.

On a microscale, β particles and other electrons follow tortuous paths in

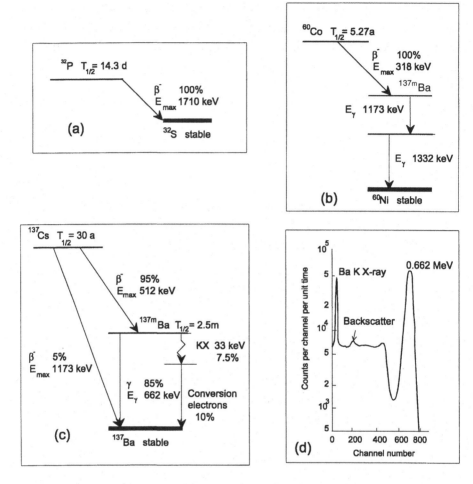

Figure 3.4. Decay schemes for three radionuclides decaying by β particle emission (see also Table 4.1). (a) Phosphorus-32, a pure β particle decay. (b) Cobalt-60, the particles are followed within 10^{-12} s by two γ rays. ^{137}Cs– ^{137m}Ba–^{137}Ba decays. (d) The photon spectrum obtained with a NaI(Tl) detector is commonly assigned to ^{137}Cs which is a pure β particle emitter as seen in (c). The spectrum shows the 662 keV γ ray peak and the average KX ray peak near 33 keV. Also shown are the Compton edge near 480 keV and the small backscatter peak near 180 keV.

matter, being continually scattered by atomic electrons as illustrated schematically in Figure 3.3(b). Nevertheless, these particles have well defined ranges that depend on the energy spectrum of the particles and on the properties of the material through which they move. The ranges are measured as the thicknesses of the material which will just stop practically all β particles of a given maximum energy (Figure 3.2(b)).

3.3.2 Beta particle applications

A large fraction of nuclear transformations are signalled by β particle emissions, followed promptly (within 10^{-12} s) by γ rays; the decay of cobalt-60 (Figure 3.4(b)) is a typical example. Most industrial applications are concerned with γ rays rather than β particles, but β particle applications are common, both in industry and the sciences (Section 7.3).

When employing β particles emitted from solid sources, it is preferable to use pure β particle emitters because there are then no competing radiations except bremsstrahlung, the effects of which can often be avoided (Section 3.8.1). Pure β particle emitters are not common. Table 8.2 in this book lists β particle emitters frequently used for industrial applications.

Two widely used pure β particle emitters of low maximum energy are carbon-14 (^{14}C, $T_{1/2} = 5730$ y, $E_{\beta max} = 157$ keV) and hydrogen-3, known as tritium (^{3}H, $T_{1/2} = 12.34$ y, $E_{\beta max} = 18.6$ keV). Their widespread use was made possible by the invention of liquid scintillation counting during the mid-1940s (Sections 4.2.4 and 5.4.2). Liquid scintillation techniques currently permit routine counting of tritium samples with reproducibilities within 1.0% (Peng, 1977, L'Annunziata, 1998, Ch. 4).

3.3.3 The scattering and backscattering of beta particles

Backscatter denotes a change in the direction of the particles from the forward to the rearward hemisphere relative to the emitter. The scattering of β particles occurs very efficiently because they are scattered by physically identical particles, the atomic electrons which abound in matter.

The intensity with which β particles are backscattered is proportional to the Z number of the backscatterer (Figure 3.5). This fact is used to monitor or identify the chemical nature of the backscatterer relative to a conveniently chosen standard (Section 7.3.2), The backscattering material must be thick enough not to transmit incident electrons, a thickness known as the saturation thickness.

When employing β particles emitted from solid sources, it is preferable to use the source making procedures to be introduced in Section 4.2.2. There are over a hundred β particle emitters routinely produced, but only a small fraction of these have the properties needed to make easily used sources for industrial and scientific applications (Table 8.2), i.e. a long enough half life, sufficient freedom from disturbing radiations and the required range of β particle energies.

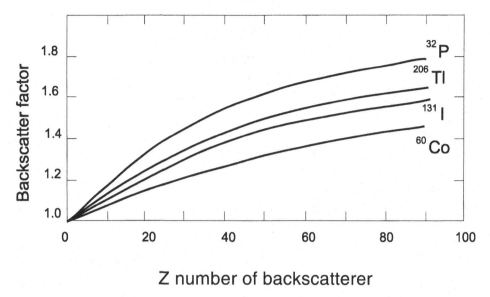

Figure 3.5. Backscatter factors (relative values), for β particles emitted from [32]P (695 keV), [206]Tl (540 keV), [131]I (182 keV), [60]Co (96 keV), as functions of the Z number of the backscatterer. The energies are the averages of each particle spectrum (based on Shapiro, 1972, Figure 4.4).

3.3.4 An introduction to beta particle spectra

Pulse height spectra due to α particles and γ and X rays will be introduced in Section 3.5. Pulse height spectra due to β particles are introduced here because the formation of these spectra differs from that of α and γ ray spectra. The preferred detectors for β particle spectrometry are crystals of silicon (Section 5.5.2). Spectrum plotting is now commonly performed employing a computerised multichannel analyser system.

Beta particle emissions signal transitions of the decaying nucleus between sharply defined energy levels (Figure 3.4). Notwithstanding this, neither negatrons nor positrons are emitted with sharply defined energies but with a spectrum of energies from zero up to a maximum (Figures 3.6(a) to (d)). This is so because the decay energy is not only carried away by β particles but shared between the β particle and the normally unobservable neutrino. The ratio between their intensities is a function of the decay scheme with no two radionuclides having the same decay scheme.

Negatrons share the decay energy with anti-neutrinos. Positrons, which are the antiparticles of negatrons, share it with neutrinos (Table 1.2). Positrons are ejected from positively charged nuclei, resulting in spectra with a relatively high average energy, such as that of sodium-22 (Figure 3.6(a)). Negatrons being

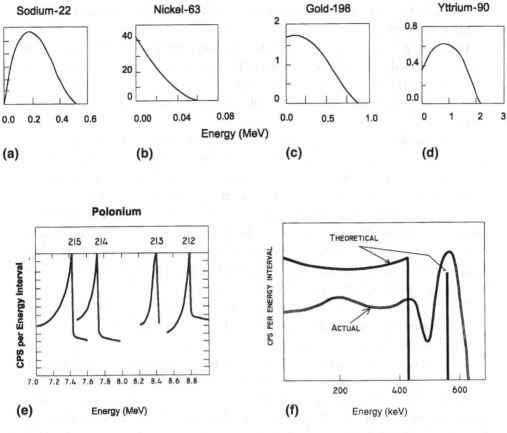

Figure 3.6.   Pulse height spectra due to β^+, β^-, α and γ rays. (a) A positron (β^+) spectrum. (b) to (d) Three negatron (β^-) spectra of nickel-63, gold-198 and yttrium-90 (Cross *et al.*, 1983). (e) Alpha particle spectra from four polonium isotopes (Mann *et al.*, 1991, Figure 4.16). (f) A 560 keV γ ray spectrum showing the theoretically calculated pulse heights (no scatter), and the observed spectrum with a shallow backscatter peak near 170 keV and the Compton edge just below 400 keV (NCRP, 1985, Figure 4.2).

emitted from a positively charged nucleus have to overcome its Coulomb attraction, with a fraction of these particles only just getting away with near zero kinetic energy as seen in Figures 3.6(b), (c), (d). (Cross *et al.*, 1983). Beta particle spectra shaped like that of ^{90}Y have the highest average energy while spectra shaped like that of ^{63}Ni have the relatively lowest average energy.

3.3.5 Surface density

The surface density of a material is defined as $t \times \rho$ g/cm^2, the product of the linear range (t cm) in that material and its physical density (ρ g/cm^3). It serves for comparing ranges of β or α particles in matter and also as a unit of γ ray attenuation (Section 3.4.4).

Assuming a beam of mono-energetic electrons, their range in a material is inversely related to its physical density. The lower the density, the longer the range. When this inverse relationship holds, the product $t \times \rho$ g/cm² can be expected to be of a similar order for all materials, whether gaseous, liquid or solid.

For example, the densities of air, water and aluminium at room temperature are approximately equal to 0.0013, 1.00 and 2.70 g/cm³ respectively. If a beam of mono-energetic electrons is absorbed in 0.10 cm aluminium, one expects it to travel 0.27 cm in water and about 208 cm in air. The measured ranges are 0.22 cm and 230 cm, a sufficiently good agreement for many purposes. Because the dimension of the unit grams per square centimetre refers to an area, the product $t \times \rho$ g/cm² was termed surface density (SD). The approximate independence of the SD range from the nature of the absorbing material applies to α as well as β particles.

This rule is convenient as a guide for nuclear radiation applications and also for radiation protection. Given the SD range of electrons of energy E in material A one can use the rule-of-thumb $(\rho \times t)_A = (\rho \times t)_B$ to predict their range in material B of known density at least approximately, and this the more accurately the closer the atomic numbers Z_A and Z_B. Ranges of electrons in the chemical elements and in a wide range of compounds were published by Seltzer and Berger (1982). They can also be obtained via the Internet (Table A3.1, Appendix 3).

3.4 Properties of gamma rays and X rays

3.4.1 Gamma rays and their decay data

High-energy electromagnetic radiations emitted from the nuclei of atoms are known as γ rays. The energies of the large majority of γ rays used for applications are between 50 keV and 3 MeV. Whereas α and β particles are frequently completely absorbed in the material through which they travel, the intensity of beams of γ rays is attenuated exponentially, when the extent of the attenuation is a useful parameter for numerous applications (Section 7.2).

Unlike α and β particles, which are emitted from parent atoms as they decay (Section 3.2.1), γ rays are follow-on radiations, being emitted from the nuclei of daughter atoms. There are other follow-on radiations besides γ rays as will be seen in Section 3.6.1. That γ rays are emitted by the daughter nuclides is easily overlooked because, following long-standing usage, they are named after the parent. One refers to cobalt-60 γ rays even though they are emitted from the daughter nickel-60, which is created with two excited states which promptly (within 10^{-12} s) de-excite in cascade (Figure 3.4(b)).

An important consequence of the fact that γ rays are emitted from the daughter nuclide relates to their multiplicity per decay. Primary decays, e.g. those signalled by β particle emissions (Sections 3.3.1), have to obey the rule of one β particle per decay. There is no such restriction on follow-on γ rays. Daughter nuclides are frequently created in numerous excited states which are then de-excited by γ rays, though de-excitations could also occur by a competing process known as internal conversion (Section 3.6.2).

When the number of excited daughter states is large, the number of de-excitations per state is invariably a small fraction of the decay rate of the parent nuclide, on average rarely exceeding the equivalent of two γ rays per parent decay. This is readily verified from decay scheme data. Multiple γ ray decay will be discussed in more detail in Section 3.7.3.

While the decay of large numbers of radionuclides result in several daughter excitations, de-excited by γ rays (see Figure 3.9(a) later), others are followed by fewer than one γ ray per decay of the parent. In Figure 3.4(c), the number of γ rays per decay of ^{137m}Ba is 90 γ rays per 100 decays and there are many much lower γ ray fractions, e.g. the 9.85 γ rays following 100 ^{51}Cr decays shown in Figure 3.11(a), about 1 gamma ray per eleven decays.

In Figure 3.4(c) the parent activity is ^{137}Cs with 95 out of 100 decays feeding ^{137m}Ba, its isomeric daughter. ^{137m}Ba de-excites partly by γ ray emissions (~90%) and partly by conversion electron emissions (~10%), a process which will be introduced in Section 3.6.2. Although the γ rays originate from ^{137m}Ba, they are known as ^{137}Cs γ rays. Their intensity relative to ^{137}Cs is ~0.95 × ~0.90 ≈ 0.85 or 85% (Figure 3.4(c), (d)).

The importance of adequate knowledge of relevant decay data of radio-nuclides employed for applications cannot be overemphasised. Unless sufficient data are known, notably the energies and intensities of the radiations of interest and their half lives, applications are unlikely to yield satisfactory results.

3.4.2 X rays

Electromagnetic radiations in the kiloelectronvolt range and higher comprise X rays as well as γ rays. Radiations are known as X rays if their energies exceed about 0.1 keV but there is no upper limit. X rays are produced as bremsstrahlung (Section 3.8.1), or as fluorescent X rays (Section 3.9.1). They are not only used in medicine but also extensively for industrial purposes and for scientific research, with fluorescent X ray spectrometry serving as a highly efficient analytical tool (Section 7.2.3).

Electromagnetic radiations carry no electric charge (Table 1.2) and so are not subject to Coulomb forces. It is here important to bear in mind that electromagnetic radiations cannot trigger detectors directly. They do so by photoelectric interaction or Compton scatter randomly knocking electrons out of atoms and transferring to them all or part of their energy which is used to excite atomic electrons as is done by β particles. Photons with an energy above 1.022 MeV also interact via pair production (see below).

3.4.3 Three types of gamma ray interactions

Figure 3.7(a) illustrates γ ray interactions as functions of γ ray energy (E_γ) and the Z number of the material in which these interactions occur, using small and large Z number examples (^{13}Al and ^{82}Pb). The figure shows three distinct and mutually independent interactions: photoelectric, Compton and pair production. Figure 4.1 in the next chapter illustrates these processes as they occur in a NaI(Tl) detector.

At low γ ray energies (<50 keV in Al, <200 keV in Pb), interactions are seen to occur principally by photoelectric processes. The γ ray gives all its energy to an outer atomic electron which is easily dislodged, when it dissipates the γ ray energy in the detector, so resulting in a pulse in the full energy peak of the spectrum (Figures 3.4(d) and 3.6(f)).

At higher γ ray energies, say >50 keV in aluminium, photoelectric interactions are increasingly replaced by inelastic Compton scatter (Figure 3.8(a)) which can occur over a wide range of energies. Following an energy loss by Compton scatter, the γ ray could escape or undergo another Compton scatter and eventually be absorbed by a photoelectric interaction. Since these processes occur well within the resolving time of the detector, they sum to again produce a pulse in the full energy peak. If the γ ray escapes (reduced in energy) the pulse due to the energies of the Compton scattered electrons will appear in the Compton continuum (Figure 3.6(f)). Other effects of these processes on γ ray spectra will be discussed in Section 3.5.3.

Pair production occurs when the γ ray energy exceeds 1022 keV and so is high enough to provide for the mass of a negatron–positron pair plus any kinetic energy. Pair production has to occur near an atomic nucleus to ensure that momentum can be conserved. It is the inverse of negatron–positron annihilation (Section 3.3.1). Although it can happen when $E_\gamma > 1022$ keV, it is seen to remain rare relative to the other types of γ ray interactions even in

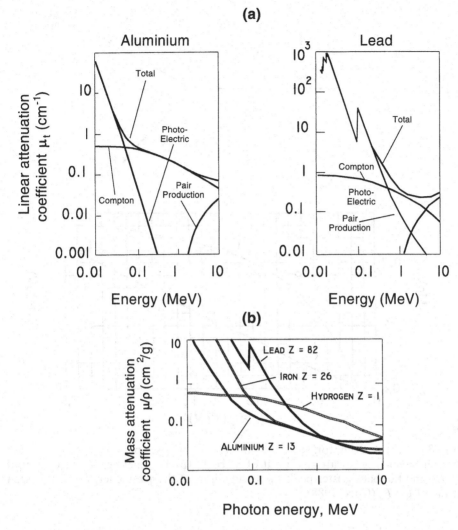

Figure 3.7. Total linear and mass attenuation coefficients (μ_t cm^{-1} and μ_m cm^2/g) for γ rays. (a) μ_t values for aluminium ($Z = 13$, $\rho = 2.70$ g/cm^3) and lead ($Z = 82$, $\rho = 11.3$ g/cm^3), showing the three mutually independent components. (b) The mass attenuation coefficient μ_m for the elements H, Al, Fe and Pb. Note the different function for hydrogen.

lead ($Z = 82$) until $E_\gamma > 2$ MeV (Figure 3.7(a)). Following pair production in sufficiently large detectors, the electron pair is normally absorbed, resulting in a full energy pulse. If one or both electrons escape, this leads to the formation of so-called single or double escape peaks to be discussed in Section 6.4.2.

Although the kinetic energy of charged particles is absorbed continuously

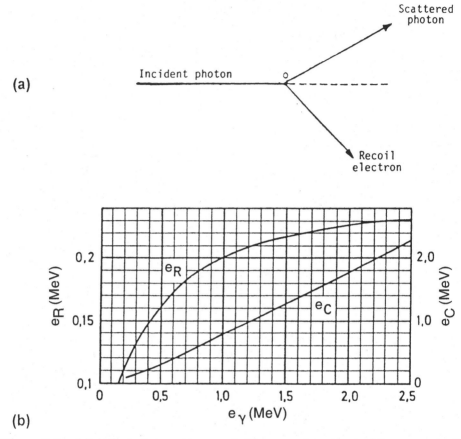

Figure 3.8.   (a) A Compton scatter event. The γ ray loses part of its energy to the atomic electron which scattered it out of its path. (b) The energies of the Compton edge E_C and the backscatter peak E_R as functions of the γ ray energy, E_γ. It is seen that $E_C + E_R = E_\gamma$ (CoN, 1988).

along their path in large numbers of individually very small steps (Section 1.3.5), γ rays can pass millions of atoms between interactions, so helping to explain the relatively high penetrating power of these rays.

3.4.4 Photon attenuation, an overview

As just noted, photon attenuation occurs as the sum of three mutually independent components: photoelectric (pe), Compton (C) and pair production (pp), though μ_{pp} is zero at γ ray energies below 1022 keV. Writing μ_t (cm^{-1}) for the total linear attenuation coefficient one has,

$$\mu_t = \mu_{pe} + \mu_C + \mu_{pp}. \tag{3.2}$$

Figure 3.7(a) shows μ_t and its three components for aluminium and lead, with μ_t depending not only on E_γ and on the Z number of the attenuator but also on its physical density.

Figure 3.7(a) shows that the E_γ and Z dependence is particularly strong for μ_{pe}. It has long been known (Bush, 1962, Section 9.7), that:

$$\mu_{pe} \propto Z^4/(E_\gamma)^3. \tag{3.3}$$

μ_{pp} is zero below 1022 keV and remains a minor component of μ_t, even when 3 MeV γ rays are attenuated in lead (Figure 3.7(a)). By contrast, the Compton component μ_C is relatively independent of both E_γ and Z. At 1 MeV, when Compton scatter is seen to dominate for all elements μ_C is about 4.5 times larger for lead than for aluminium, which is closely equal to the ratio of their physical densities.

Gamma ray applications are often in the range of energies where Compton scatter is dominant. With μ_t (cm^{-1}) being approximately proportional to ρ (g/cm^3), introduction of the mass attenuation coefficient μ_m followed defined as $\mu_m = (\mu_t/\rho)$ cm^2/g (Figure 3.7(b)). The coefficient μ_m is then almost independent of ρ, which is particularly useful in the Compton range of energies as will be seen in Section 3.4.6.

3.4.5 Attenuation equations for narrow beam geometry

Narrow beam geometry has to be mentioned here even though details will be introduced in the next chapter.

Whenever practical, γ rays emitted from a source should be collimated, as shown in Figure 4.6(a), a procedure known as narrow beam geometry. This is employed to prevent γ rays which lose energy during Compton scatter (Figure 3.8(a)) from reaching the detector, so reducing the peak-to-total ratio. If there is no collimation, one refers to broad beam geometry (Figure 4.6(b)), the additional number of detected γ rays being known as build-up (Figure 4.6(c)). Since the extent of build-up is difficult to predict γ ray attenuation should, whenever possible, be carried out in narrow beam geometry and this will be assumed in what follows unless otherwise stated. There are situations, notably during γ ray applications in open country, when build-up is difficult to prevent or could even be an asset (Section 8.2.4).

Given narrow beam geometry and a total linear attenuation coefficient μ_t, gamma rays lose their intensity I (measured in gamma rays per square centimetre per second, γ/cm^2/s) in accordance with:

$$-dI/dt = \mu_t \times I, \tag{3.4}$$

where dt is a small distance in the material and μ_t depends on the three components shown in Eq. (3.2). Hence:

$$I_t = I_0 \exp(-\mu_t \times t), \tag{3.5}$$

where I_0 is the intensity of the unattenuated beam and I_t its intensity after travelling t cm in the material of interest. Equation (3.5) is employed for all attenuation calculations that rely on μ_t or any of its components.

Photon beams employed for applications frequently have to pass through a number of different materials before reaching the detector, each with its own μ_t and Δt. Writing Eq. (3.5) as:

$$\ln(I_0/I_t) = \mu_t \Delta t \tag{3.6}$$

the term $\mu_t \Delta t$ is expanded as required, say to i terms, so obtaining:

$$\ln(I_0/I_t) = \mu_{t_1} \times \Delta t_1 + \mu_{t_2} \times \Delta t_2 + \cdots + \mu_{t_i} \times \Delta t_i. \tag{3.7}$$

These equations apply for narrow beam geometry.

The magnitude of attenuations is conveniently estimated with the help of the half thickness $t_{1/2}$. On substituting $I_t = 0.5 I_0$ in Eq. (3.5), one obtains:

$$t_{1/2} = \ln 2/\mu_t = 0.693/\mu_t, \tag{3.8}$$

with $t_{1/2}$ values depending on the same variables as μ_t. Half value thicknesses for three commonly used materials are shown in Table 7.3.

3.4.6 Photon attenuation measurements using μ_m

When replacing μ_t (cm^{-1}) by μ_m (cm^2/g) (Section 3.4.4), it is necessary to replace the linear distance t cm by the surface density SD (g/cm^2) (Section 3.3.5) to ensure a dimension-free exponent. Equation (3.5) then becomes:

$$I_{SD} = I_0 \exp(-\mu_m \times SD). \tag{3.9}$$

Figure 3.7(b) shows that μ_m values are very similar for all light elements ($Z < 35$) at γ ray energies between about 0.5 and 2.0 MeV and for nearly all stable elements for γ ray energies between 1 and 2 MeV.

Working with, say 1.33 MeV γ rays from ^{60}Co and with metal strips which are identical in thickness, any fault in their internal structure which affects their physical density, e.g. a void due to a faulty weld, can be identified from differences in the attenuation of the γ rays relative to that in a suitably chosen standard. This is so because the γ rays that traversed an air space (void) within the metal travel a different distance in SD terms than the rays that did not encounter voids. This effect is also the basis of γ ray radiography to be discussed in Section 7.2.1.

Although μ_m is almost independent of Z between about 0.5 and 2 MeV, this is not so below 0.1 MeV (Figure 3.7(b)) which makes the attenuation of low-energy γ or X rays a sensitive monitor, e.g. for the detection of chemical impurities in foils of metals or plastics. Accurate attenuation measurements at low photon energies can only be made on sufficiently thin samples.

Only hydrogen ($Z = 1$) does not conform to the general pattern (Figure 3.7(b)). The marked difference between μ_m for hydrogen and for other molecules with significant hydrogen content relative to hydrogen-free materials is extensively employed in many applications.

3.5 Pulse height spectra due to alpha particles and gamma rays

3.5.1 The response of detectors

Alpha and beta particle emissions signal nuclear decays between sharply defined energy levels belonging respectively to the parent and daughter nuclide. Why the interaction of β^- or β^+ particles in energy-sensitive detectors fails to produce peaks was stated in Section 3.3.4. However, α particles carry away the entire energy liberated by an α particle emitting decay, e.g. the decay of radium-226 to radon-222 (Figure 1.5), except for recoil energies left with the emitting atoms. If all α particles from a source reach the detector without losing energy between source and detector (Section 3.2.2), they give rise to pulses of identical height (except for random fluctuations), so yielding a narrow full energy peak (Section 3.5.2).

The situation is more complex for γ rays. Most nuclear decays occur via several excited states commonly de-excited by γ rays (but also in part by conversion electrons, see Section 3.6.2) leading to multi-energy γ ray spectra (Section 3.4.1). Gamma ray detectors should be able to resolve peaks due to closely spaced γ ray energies which is effectively done by high resolution semiconducting detectors (Figure 3.9(b),(c)).

Although the energy resolution of NaI(Tl) detectors at room temperature is 20 to 40 times lower than that of germanium detectors at liquid air temperature (Figure 3.10), the former have the advantage of being considerably less expensive for a comparable size, plus they are easier and cheaper to operate.

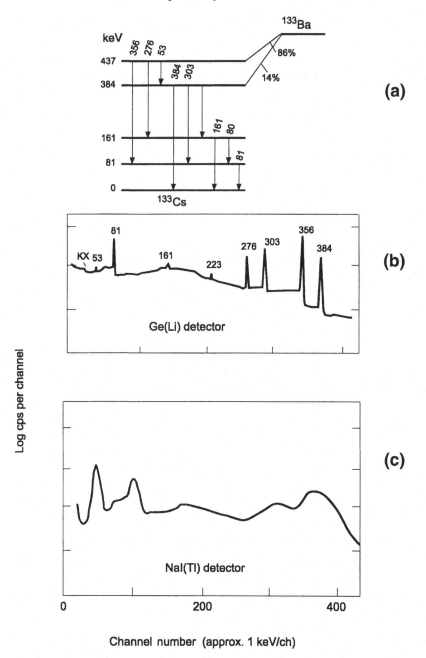

Figure 3.9. (a) The decay of barium-133 ($T_{1/2} = 10.6$ y) by electron capture, followed by nine γ rays. (b) The ^{133}Ba spectrum obtained with a Ge(Li) detector, and (c) with a NaI(Tl) detector (Debertin and Helmer, 1988, Figure 4.25). The latter detector clearly displays low-energy photons which are barely shown by the Ge(Li) detector. Graphs (b) and (c) have the same energy scales.

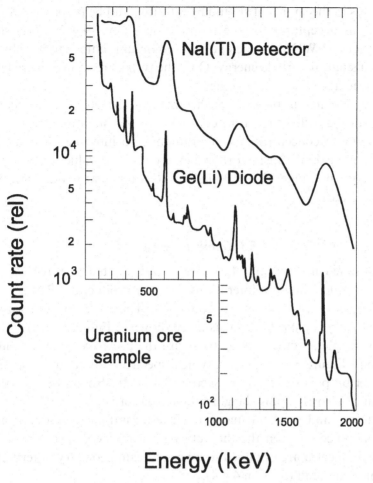

Figure 3.10.   A comparison of γ ray spectra due to a uranium ore sample obtained with a Ge(Li) detector and a NaI(Tl) detector.

3.5.2 *Alpha particle spectra*

The preferred method for obtaining α particle spectra makes use of ion implanted silicon detectors (Section 5.5.4). These are small crystals mounted in chambers which can be evacuated and which are fitted to permit accurately reproducible source–detector geometry. The crystals range in thickness from about 0.1 mm upwards to several millimetres.

The range of 10 MeV α particles in silicon is only about 0.1 mm so that a 0.1 mm or at most a 0.3 mm thick silicon detector, with source and detector mounted in vacuum, satisfies requirements for spectrometry when employing α particles from naturally occurring radionuclides (Figure 1.5). To minimise energy losses in the source, it should be ultrathin and be prepared with as low an activity as acceptable for the required accuracy. Figure 3.6(e) shows α particle spectra (FWHM~15 keV) due to polonium radioisotopes which each emit only a single α particle energy. Only the peaks are shown, there being no identifiable features at lower energies.

When employing α particle spectrometry routinely, e.g. to identify α particle emitting pollutants from coal fired power stations which normally emit measurable concentrations of uranium, thorium and their daughters, this can often be done satisfactorily when the FWHM values of the α particle spectra are in the range 50 to 100 keV. The equipment is then cheaper and easier to operate.

3.5.3 Gamma ray spectra

As was outlined in Section 3.4.3, γ rays are detected in three ways – via photoelectric and Compton interations and pair production. Pair production will be further dealt with in Section 6.4.2. The other two processes are responsible for γ ray spectra such as those shown in Figures 3.4(d) and 3.6(f). Relatively simple spectra such as these are the preferred type for industrial applications. Figure 3.6(f) shows a typical spectrum for gamma rays that did not interact by pair production (Section 3.4.3). It also shows the effects of random fluctuations in pulse heights due to the statistics of γ ray interactions in the detector and adjacent materials. These random interactions account for the difference between the theoretical and observed spectrum shown in the figure. If spectra are obtained in a broad beam geometry, interpretations become more difficult (Section 4.4.4).

The multiple interactions of γ rays even when emitted at identical energies results in a pulse height distribution from near zero energy to a maximum exceeding the emitted γ ray energy due to the quasi-Gaussian shape of the full energy peak shown in the figure. The fraction of pulses reaching the full energy peak depends principally on the energy with which the γ rays were emitted and the conditions prevailing during the measurement. The ratios of pulses in the peak to pulses in the entire spectrum (peak-to-total ratio) obtained in optimum conditions as functions of the γ ray energy and the dimensions of NaI(Tl) detectors are shown in Figure 4.5 in the next chapter.

Coming to the near zero and low-energy pulses shown e.g. in Figure 3.6(f), they are due predominantly to photoelectric absorption of γ rays which had been scattered to low energies (Figure 3.7(a)). On the other hand, the energies of the pulses due to Compton scatter for a given initial energy extend from low values upwards to the so-called Compton edge. Here again, the statistics of interactions cause the observed spectrum to show broad rather than sharp edges. The Compton edge is due to electrons dislodged when γ rays are fully backscattered within the detector and adjacent material (Figure 3.8(b)).

Further details about the formation of γ ray spectra go beyond the scope of this book, except as regards backscattered γ rays (150° to 180°) lead to pulses in the so-called backscatter peak shown in Figure 3.4(d). This peak is small as is the backscatter peak in Figure 3.6(f) since these spectra were obtained in a relatively scatter-free environment. However, these peaks can become dominant in spectra due to strongly scattered radiations (Section 3.9.1).

The energies of the Compton edge (E_C) or the backscatter peak (E_B) are not always easily identified, especially so in multi gamma ray spectra. However, if the position of one of them is known, that of the other can be calculated since $E_C + E_B = E_\gamma$, the energy of the interacting γ ray (Figure 3.8(b)).

3.6 Electron capture (EC), gamma rays and conversion electrons

3.6.1 EC decays and their use as quasi-pure gamma ray emitters

Electron capture (EC) decays occur (Figure 3.11) when an atomic electron is pulled into the nucleus of its atom where it can be pictured to turn a proton into a neutron. EC decays are the inverse of β^- decays. An EC decay is a primary transformation, decreasing the Z number of the parent by one unit (see Figure 3.11) without the emission of primary radiations (except for neutrinos), though there are follow-on radiations emitted by the daughter nuclide.

Between 80 and 90% of EC decays originate in the K shell of their atoms, the shell nearest to the nucleus. The resultant vacancies are at once filled by electrons from beyond the K shell. The energy liberated during these transfers is emitted mainly either as KX rays or as K-Auger electrons (Figure 3.12).

Each electron transfer from an outer to an inner shell leads to the release of binding energy which appears either as a fluorescent X ray (Section 3.9.1) or as the kinetic energy of an Auger electron. The ratio of the intensities of fluorescent X rays to those of Auger electrons is known as the fluorescent yield, denoted ω, which is a function of the atomic number Z of the element of interest (Section 3.9.3). For each element fluorescent X ray energies are significantly high for interactions in the K shell (ω_K), but for outer shells they are commonly too low to be measured except by specialists. The dependence of ω_K on Z is illustrated in Figure 3.12.

The decay energy liberated during the EC process is carried away by ordinarily unobservable mono-energetic neutrinos, each neutrino carrying

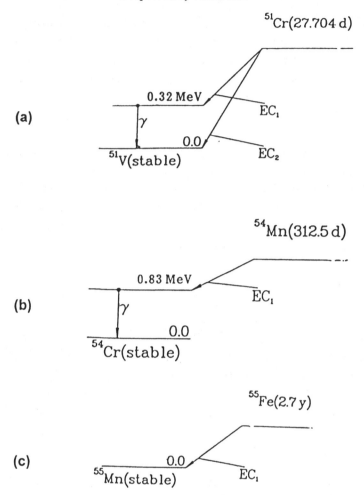

Figure 3.11.   Decay schemes of three electron capture decays followed (for two of them) by a single γ ray. (a) Chromium-51 to vanadium-51 ($E_\gamma = 320$ keV, $f_\gamma = 9.86\%$). (b) Manganese-54 ($E_\gamma = 835$ keV, $f_\gamma = 99.98\%$). (c) Iron-55 (no gammas). Based on Lagoutine *et al.* (1978).

away all the decay energy and not only part of it as happens during decays by β particle emissions (Section 3.3.4).

For a few radionuclides decaying by EC, decays go direct to the ground state of the daughter. However, as a rule decays proceed to one or more excited states. If there is only a single excited state, de-excited by no more than one or two γ rays with other follow-on radiations being of much lower energy, the radionuclide could be labelled a quasi-pure γ ray emitter.

Figure 3.11 shows the decay schemes for the EC decays of chromium-51 and manganese-54, two examples of quasi-pure γ ray emitters which are often

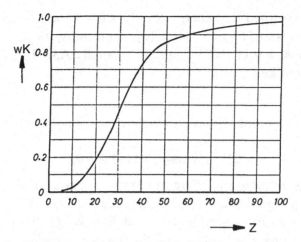

Figure 3.12.   The fluorescent yields in the K shell of atoms (ω_K) as a function of the Z number of the atoms (Mann *et al.*, 1991, Section 2.4.5).

useful for applications. The low-energy KX rays (<6 keV) and the K Auger electrons (<5 keV) from these nuclides are unlikely to interfere with the use of the much higher energy 0.32 and 0.83 MeV γ rays.

3.6.2 *The internal conversion process*

The de-excitation of daughter nuclides does not only proceed via γ rays but also by a competing process known as internal conversion (IC). Each contributes a share, but the shares are commonly different (see below). During IC, the de-excitation energy is transferred from the nucleus of the daughter to an inner shell atomic electron, usually a K electron which is nearest to the nucleus. The electron is ejected from its shell, dissipating the energy received from the de-excitation, except for the part used to liberate it from its atom (NCRP, 1985, Appendix A:1.2.4). Both EC and IC involve the nucleus of the atom and lead to the ejection of inner shell atomic electrons. Also, both lead to the same follow-on radiations, fluorescent X rays and Auger electrons.

Internal conversion rates relative to the rate of γ ray emissions are calculated using the total internal conversion coefficient a_t. One has:

$$a_t = N_{ce}/N_\gamma, \tag{3.10}$$

where N_{ce} stands for the fraction of de-excitations via conversion electrons (between 80 and 90% are emitted from the K shell) and N_γ represents the fraction proceeding via γ rays. Although there are exceptions, a_t is less than 0.01 when the de-excitation energy exceeds about 200 keV, but it could exceed a hundred when that energy is only a few kiloelectronvolts. For the ^{137m}Ba decay shown in Figures 3.4(c), (d), the ratio N_{ce}/N_γ is approximately equal to $0.10/0.90 \approx 0.11$ (NCRP, 1985, Appendix A3). This is a relatively high IC rate at 662 keV.

3.7 The role of mass energy in determining nuclear decays

3.7.1 Neutron-poor radionuclides

Radioactivity is triggered by differences in the energy equivalent of the mass of neighbouring nuclides in appropriate circumstances. The energy is expressed in kilo- or megaelectronvolts and labelled Q. It provides for all requirements during the decay, e.g. the kinetic energies of the emitted radiations both primary and follow-on, as well as the energies required to eject electrons from their atoms.

There is also the binding energy holding protons and neutrons together in their nuclei (Figure 1.2), and the energy equivalent of the mass of each emitted particle. This is about 3730 MeV for α particles (Table 1.2), but only 0.511 MeV for β particles, about one part in 7300 of that of α particles. It is the low-energy equivalent of the mass of electrons which contributes to the large number of β decays relative to α decays.

Figure 1.3 shows the stable nuclides near the middle of each row with the neutron-poor radionuclides on the left and the neutron-rich radionuclides on the right. Neutron-poor radionuclides have two pathways towards increased or complete stability – EC decay and decay by positron emissions. If EC decays occur to the ground state of the daughter, the fluorescent X rays and Auger electrons are the only emitted radiations. If the EC decay is to an excited state de-excited by γ radiations which are unconverted or nearly so, one has γ rays in addition to the X rays and Auger electrons. If IC is also present, it reduces the γ ray fraction with a corresponding increase in the fraction of conversion electrons.

Iron-55 decays direct to the ground state of its stable daughter, manganese-55 (Figure 3.11(c)). Because there are no emissions at higher energies, the averagely 6 keV manganese KX rays are available, e.g. for the calibration of low-energy photon detectors.

3.7.2 Positron decay and positron tomography

When β^- or EC decays occur direct to the ground state of the respective daughters, the Q values applying to β^- decays have to provide for the mass

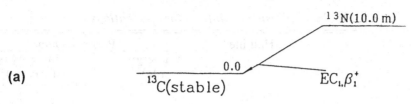

(a)

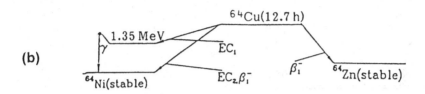

(b)

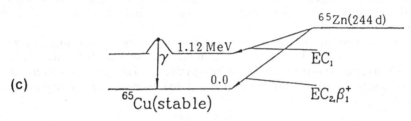

(c)

Figure 3.13.   Three decay schemes illustrating different decay modes. (a) The decay of nitrogen-13: 99.8% β⁺, 0.2% EC. (b) The three-pronged decay of copper-64: 18% β⁺, 37% β⁻, 45% EC. (c) The decay of zinc-65: 98.5% EC, 1.5% β⁺.

and kinetic energies of the particles plus that of their anti-neutrinos and the binding energy, but that is all. For pure EC decays, e.g. that of ^{55}Fe, they have to provide for the neutrino energies and the binding energies needed to dislodge the electrons. However, positron decay inevitably leads to positron–negatron annihilation (Section 3.3.1) and so cannot compete with EC decay unless there is sufficient difference in nuclear mass (and so in energy) to provide for the annihilation events which always follow positron emissions.

Positron decays are invariably accompanied by at least a small percentage of EC decays, 0.2% for carbon-11 and nitrogen-13. However, EC percentages can be large (e.g. 95% for zinc-65 (Figure 3.13)). As a rule, the 511 keV annihilation γ rays are more useful for applications than β⁺ particles. Positron decays are gaining in importance on account of their role in computerised tomography (Section 7.2.1). Because they can be used to provide three dimensional imaging (Section 1.4.4), tomographic techniques

Table 3.1. *Positron emitters for applications.*[a]

Radionuclide	Half life[b]	Positron Data	
		Energy (MeV)	Intensity per decay (%)[c]
Carbon-11	20.4 m	0.97	99.8
Nitrogen-13	9.97 m	1.20	99.8
Oxygen-15	2.04 m	1.70	99.9
Fluorine-18	109.7 m	0.64	97.0
Sodium-22	2.60 y	0.55	89.8
Aluminium-26	7.3×10^5 y	1.17	81.8
Vanadium-48	16.0 d	0.70	50.3
Cobalt-58	70.82 d	0.48	15.0
Copper-64	12.70 h	0.65	17.9
Gallium-68	68.0 m	0.84[d]	88.0

Note: [a] The γ ray emissions (if any), following the decay of the positron emitters included in this table, are unlikely to interfere with the employment of annihilation γ rays.
[b] Positron emitters which decay with very short half lives can only be used for applications if there is a direct access to a cyclotron and associated radiochemical facilities.
[c] Positron decays are invariably accompanied by EC decays and sometimes also by negatron decays (Section 3.6.1), so accounting for the less than 100% intensities.
[d] This is the average positron energy for two branches. The other nine radionuclides decay via a single positron branch and the stated energies are the maxima of the spectra.

are being increasingly applied to industrial processes e.g. in the construction industry for inspecting castings to detect cracks, or the study of flow patterns in pneumatic pipelines.

Tomographic imaging producing displays in three dimensions requires the processing of very large amounts of data and so has to be computer assisted (CAT). Table 3.1 lists positron emitters which could be useful for such applications but industrial tomography is more commonly employed with X rays (Section 7.2.3) than with the 511 keV γ rays following positron emissions.

3.7.3 Multi gamma ray emitters

The introduction and development of NaI(Tl) detectors and the associated signal processing equipment from the mid-1940s onwards made it possible to record multi gamma ray spectra with usefully high resolution and this almost as a matter of routine (Crouthamel 1960, 1981, Heath 1964). Then came another very big step forward: the advent of high-resolution semiconductor γ

and X ray detectors beginning during the mid-1960s. High-resolution semi-conductor γ ray detectors are either crystals of lithium drifted germanium (Ge(Li)) or high-purity germanium (HPGe) which make it possible to routinely obtain spectra such as that of europium-152 (Figure 3.14(a)).

Europium-152 ($T_{1/2} = 13.51$ y) undergoes a two-pronged decay (EC and β⁻) followed by the emission of γ rays with over 100 different energies between 0.12 and 1.53 MeV, but the overall γ ray emission rate is only about 1.5 γ rays (of all energies) per decay of ^{152}Eu. The large majority of these energies can be clearly resolved although only eleven energies are emitted with sufficient intensity (>3% each) to be useful for intensity calibrations. The remaining nearly 90 energies are emitted with, on average less than 0.03% probability each, so explaining the many very small peaks seen in Figure 3.14(a).

There are many other effects which could cause distortions in gamma ray spectra, notably Compton backscatter. This is illustrated in Figure 3.15, with details to come in Section 3.9.1.

Multi gamma ray emitters are of particular interest for energy and intensity calibrations of high-resolution detectors when it is often possible to effect calibrations over the entire energy range of interest using a single multi γ ray emitter, with europium-152 as an outstanding example. Other long-lived multi γ ray emitters frequently used for detector calibrations are europium-154 and, for the lower energy range, barium-133 (Table 8.1). It is desirable to use long half life radionuclides for this purpose, to justify the work involved in the calibration. By contrast, NaI(Tl) detectors are rarely able to adequately resolve more than a few of the peaks of multi γ ray emitters (Figure 3.9(c)) except when used by specialists.

The lower energy section of spectra due to multi gamma ray sources is invariably positioned on top of a large background due mainly to Compton scatter. Pulses due to γ rays which were Compton scattered out of the high energy peaks build up below the lower energy part of the spectrum (Figures 3.10, 3.14(a) and 3.15). Background subtractions to obtain the net count in full energy peaks are readily made for high-energy peaks but require considerable care for the low-energy peaks since there are often many more pulses in the background just below the peak than in the peak. This applies to germanium detectors as much as to NaI(Tl) detectors.

For reasons yet to be described (Section 5.5.2), lithium drifted germanium detectors cannot be prepared without dead layers which absorb all γ rays with energies below 30 keV (or sometimes below 40 keV) and cause significant attenuation of higher energy γ rays up to about 100 keV. This is evident in Figures 3.14(c) and 3.14(b) which respectively show the contrast between a

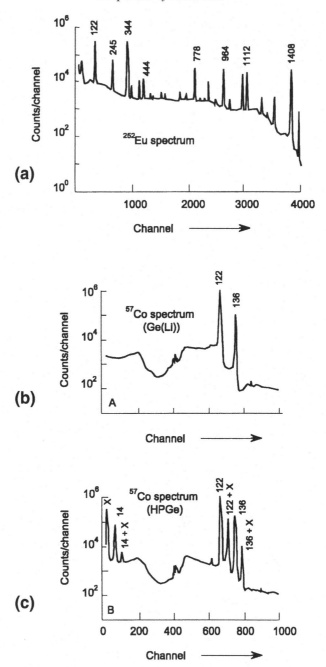

Figure 3.14.   (a) The γ ray spectrum following ^{152}Eu decays obtained with a Ge(Li) detector. Note the numerous small peaks due to very low-intensity γ rays (Section 3.7.3). (b) ^{57}Co spectra obtained with a Ge(Li) and (c) with a HPGe detector. Note the effect of the dead layer coating the Ge(Li) detector (Debertin and Helmer, 1988, Figures 4.3, 4.26).

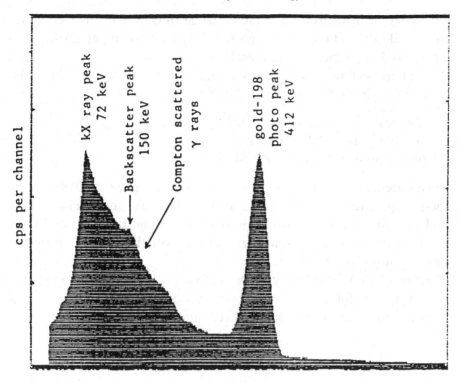

Energy (keV)

Figure 3.15. A pulse height spectrum obtained with a 50 mm × 50 mm NaI(Tl) detector of intensely backscattered γ rays from gold-198. The 411.8 keV full energy peak, dominant in narrow beam geometry (see figure 4.5), is 'dwarfed' by pulses due to Compton scatter. Two low-intensity higher energy γ rays are not shown.

[57]Co spectrum obtained with a high-purity germanium detector and the lithium drifted type. Ge(Li) detectors are now being replaced by HPGe detectors whenever practical (Sections 5.5.2 and 5.5.3).

The calibration and use of multi-gamma ray emitters requires specialist equipment and skills. Readers are referred to Debertin and Helmer (1988), Sections 3.4 and 4.1 or L'Annunziata, Ed. (1998, Ch. 3 by Fettweis and Schwenn), for a full treatment of this subject.

3.7.4 Three-pronged decays

A few radionuclides decay via three primary branches when they provide examples of the role of mass energy in nuclear decay. One such radionuclide which has found many applications is copper-64. Figure 3.13 shows examples of a β^+ decay of a neutron-poor radionuclide (nitrogen-13) as well as the three-pronged decay of copper-64 and the EC plus β^+ decay of zinc-65. It is the copper-64 decay which is of interest here.

Tables of nuclear masses quoted by Mann *et al.* (1980, Ch. 5, see also Wapstra and Audi, 1985), provide the following data.

Nickel-64: nuclear mass, 63.9280 amu; $Z = 28$
Copper-64: nuclear mass, 63.9298 amu; $Z = 29$
Zinc-64: nuclear mass, 63.9292 amu; $Z = 30$.

The mass difference for the decay of copper-64 to nickel-64, 0.0018 amu has the energy equivalent of 0.0018×932 MeV $= 1.68$ MeV, so permitting both EC and positron decay. Allowing for some rounding, the mass difference copper-64 to zinc-64 is 0.0006 amu and provides for the 576 keV maximum energy of the negatron spectrum.

Copper-64 emits 511 keV γ rays with a conveniently short half life (12.7 h), which makes it useful for tracing copper and other metals while avoiding problems due to the disposal of longer-lived radioactive residues.

3.8 Bremsstrahlung

3.8.1 *Its origin*

Bremsstrahlung (from the German for braking radiation) is only indirectly related to nuclear emissions. It is generated when β particles or other electrons emitted from radionuclides or from other sources are slowed down in matter.

When charged particles are decelerated (braked) in the electric field surrounding the nuclei of atoms, a fraction or all of the kinetic energy of the particles (depending on the type and thickness of the material), is converted into electromagnetic radiation, the bremsstrahlung. Electrons are light enough to be sufficiently strongly decelerated to produce bremsstrahlung photons of significant intensities. The relatively very much heavier protons and α particles (Table 1.2), are subject to proportionately smaller deceleration when the bremsstrahlung yield can commonly be ignored unless one is dealing with high activities.

3.8.2 *Bremsstrahlung intensities*

When emitted from their atoms, the energies of β particles range from near zero to a maximum value which depends on the radionuclide (Figure 3.6(a) to (d)). Endpoint energies are as small as 18.6 keV (^{3}H), but can exceed 2 MeV (Table 8.2). Bremsstrahlung photons due to the braking of these β particles cover the same range of energies though their average energy is nearer to $0.25E_{\beta max}$ than to the $0.33E_{\beta max}$ which applies for β particle emitters (Wilson, 1966).

The bremsstrahlung due to a high-activity β particle source is capable of generating intense ionisations, especially so if the braking occurs in high atomic number material (Eq. 3.11). If so, the bremsstrahlung intensity has to be monitored to ensure adequate protection for experimenters or to avoid errors when radioactivity measurements make use of ionisation processes.

If β particles or other electrons move a small distance dx in a material of atomic number Z they lose kinetic energy $-dE$ through collisions causing excitations and ionisations of atomic electrons $(-dE/dx)_{coll}$ and through the conversion of kinetic energy into bremsstrahlung $(-dE/dx)_{rad}$. From comments in NCRP 1985 (Appendix A1.2.9), one has approximately:

$$(-dE/dx)_{rad}/(-dE/dx)_{coll} \approx E_{\beta av} \times Z/700, \tag{3.11}$$

where $E_{\beta av}$ stands for the average energy of the β particle spectrum expressed in megaelectronvolts. If ^{32}P particles ($E_{\beta av} = 0.70$ MeV) are absorbed in copper ($Z = 29$), the fraction of the initial kinetic energy of the β particles converted to bremsstrahlung is $0.70 \times 29/700 = 0.03$ or about 3%.

The X rays discovered by Röntgen were bremsstrahlung photons produced when electrons emitted from a hot cathode in an evacuated glass tube and subsequently accelerated by an electric field were stopped in metal parts and in the glass envelope of the vacuum tube. The high-energy, high-intensity X rays required for modern medical and industrial purposes (well up in the megaelectronvolt range) are produced by accelerating electrons in high-voltage machines and decelerating them in the first few millimetres of high Z number water-cooled targets, commonly tungsten targets ($Z = 74$).

3.9 Fluorescent radiations

3.9.1 Fluorescent X rays

Fluorescent X rays and Auger electrons (Section 3.9.3) are emitted as a consequence of the transfer of electrons from outer to inner shells of their atoms (Sections 3.6.1 and 3.6.2). These radiations carry away the energy that is liberated during the transfer. The measurement of fluorescent X ray energies is a long practised technique for chemical analysis because these

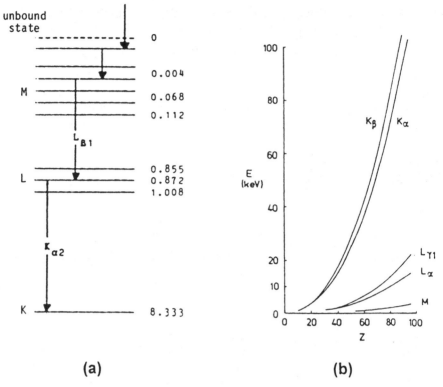

(a)   **(b)**

Figure 3.16.   (a) A typical series of de-excitations in atoms of nickel ($Z = 28$), showing the energies of the K. L, M electronic shells. (b) The energies of K, L, M X rays as functions of the atomic number of the emitting atoms (Debertin and Helmer, 1988, Figures 1.4 and 1.7).

energies are functions of the Z number of the element from which they were emitted (see e.g. Mann *et al.*, 1991, Section 2.4.5).

As explained in physics texts (see e.g. Mann *et al.*, 1980, p. 60), electrons are held in their shells by accurately measurable binding energies which decrease from inner to outer shells (Figure 3.16(a)). If a vacancy occurs in the K shell it is promptly filled by electrons from an outer shell, when the difference in the binding energies is liberated as fluorescent X rays or Auger electrons. The X rays are labelled fluorescent since they persist no longer than the excitations (10^{-12} s).

Quantum theory demands that the K shell of atoms can only accept atomic electrons for the first two elements: hydrogen and helium (Figure 1.1). When $Z > 2$, the filling of the K shell is followed by that of the L shell, then M, N, etc, with the order in which electrons are added to the shells depending on rules derived from quantum theory. The shell and subshell structure shown in Figure 3.16(a) applies to nickel ($Z = 28$).

Figure 3.16(b) shows how the KX ray energies increase with increasing Z number. The X rays labelled $K_{\alpha1}$, $K_{\alpha2}$ and $K_{\beta1}$ are not only the most energetic but also the relatively most abundant for each element. This is so because the large majority of vacancies caused by radioactive decays occur in the K shell of the atom which is nearest to the decaying nucleus.

For example, the filling of vacancies caused by electron capture in the K shell of barium atoms ($Z = 56$) leads to the emission of KX rays with the range of energies shown in Table 3.2. These energies can be resolved using high resolution detectors. NaI(Tl) detectors can only display a peak representing the average energy which in this case is close to 33 keV.

Figure 3.15 shows the average energy (~72 keV) of the KX ray peak due to mercury-198 (daughter of gold-198), obtained with a 50×50 mm NaI(Tl) detector. It is included here as an example of a fluorescent KX ray peak competing against a large Compton backscatter peak and the full energy peak due to the approximately 96% of ^{198}Au decays a prominently displayed peak when the γ rays are detected in narrow beam geometry. High-intensity Compton scatter as shown in Figure 3.15 is not uncommon during industrial applications when the structure of the irradiated material causes γ ray scatter prior to detection (Section 4.4.4).

3.9.2 *Inner shell transitions*

Figure 3.16(a) shows that for atoms of nickel ($Z = 28$), electron transitions leading to X rays of readily detectable energies occur only from the L shell to the K shell, known as inner shell transitions. By contrast X rays generated in atoms of mercury ($Z = 80$) involve transitions between seven electronic shells, though only the inner shell transitions to the K shell contribute significantly to the average energy of the about 72 keV KX rays.

Assuming one works with an element (of atomic number Z), emitting KX ray energies within a few kiloelectronvolts of 30 keV which are detected with a NaI(Tl) crystal, one can predict the average energy of these KX rays as:

$$(E_{KXav})Z = (E_{BK} - E_{BL})Z, \tag{3.12}$$

where E_{BK} is the average binding energy of electrons in the K shell of element Z and E_{BL} that of electrons in the L shell.

For the barium atoms ($Z = 56$) generated by the decay of caesium-137 (Figure 3.4(c), (d)), the average binding energies are, in rounded figures, near 38 keV in the K shell and near 5 keV in the L shell (Table 3.2). The average KX ray energy for barium is then given by:

Table 3.2. *Atomic electron binding energies and fluorescent X ray energies for the atoms of selected elements.*[a]

Z	Element[b]	Atomic electron binding energies in stated shells (keV)				X ray energies (keV) due to electron transfers to the K shell and their weighted average			
		K	L_2	L_3	M_3	$K–L_2$	$K–L_3$	$K–M_3$	$(E_{KX})_{av}$[c]
20	Ca	4.04	0.35	0.35	0.03	3.69	3.69	4.01	3.7
21	Sc	4.49	0.41	0.40	0.03	4.08	4.09	4.46	4.1
22	Ti	4.97	0.46	0.46	0.03	4.51	4.51	4.94	4.6
23	V	5.47	0.52	0.51	0.04	4.95	4.96	5.43	5.0
24	Cr	5.99	0.58	0.57	0.04	5.41	5.42	5.95	5.5
25	Mn	6.54	0.65	0.64	0.05	5.89	5.90	6.50	6.0
26	Fe	7.11	0.72	0.71	0.05	6.39	6.40	7.06	6.5
27	Co	7.71	0.79	0.78	0.06	6.92	6.93	7.65	7.0
28	Ni	8.33	0.87	0.85	0.07	7.46	7.48	8.26	7.6
29	Cu	8.98	0.95	0.93	0.07	8.03	8.05	8.91	8.2
30	Zn	9.66	1.04	1.02	0.09	8.62	8.64	9.57	8.7
35	Br	13.47	1.60	1.55	0.18	11.88	11.92	13.30	12.4
40	Zr	18.00	2.31	2.22	0.33	15.69	15.78	17.67	16.0
45	Rh	23.22	3.15	3.00	0.50	20.07	20.22	22.72	20.9
50	Sn	29.20	4.16	3.93	0.71	25.04	25.27	28.49	25.8
55	Cs	35.99	5.36	5.01	1.00	30.63	30.98	34.99	31.7
56	Ba	37.44	5.62	5.25	1.06	31.82	32.19	36.38	32.9
60	Nd	43.57	6.72	6.21	1.30	36.85	37.36	42.27	38.2
70	Yb	61.33	9.98	8.94	1.95	51.35	52.39	59.38	53.6
73	Ta	67.42	11.14	9.88	2.19	56.28	57.54	65.23	58.9
74	W	69.52	11.54	10.21	2.28	57.98	59.31	67.24	60.7
79	Au	80.72	13.73	11.92	2.74	66.99	68.80	77.99	70.5
80	Hg	83.10	14.21	12.28	2.85	68.89	70.82	80.25	72.5
83	Bi	90.53	15.71	13.42	3.18	74.82	77.11	87.35	79.4

| 90 | Th | 109.65 | 19.69 | 16.30 | 4.05 | 89.96 | 93.35 | 105.60 | 95.6 |
| 92 | U | 115.60 | 20.95 | 17.17 | 4.30 | 94.65 | 98.43 | 111.30 | 100.8 |

Note: [a] These data are based on Firestone (1987) with $(E_{KX})_{av}$ weighted using abundances normalised to 100 for the (K−L$_3$) line (K$_{\alpha 1}$).

[b] The table includes energies for selected elements between $Z = 20$ and $Z = 83$ (the highest Z number for stable elements) and also for thorium ($Z = 90$) and uranium ($Z = 92$). X ray energies due to elements with $Z < 20$ are rarely high enough for applications. X ray energies for omitted Z numbers below $Z = 92$ can be estimated to the nearest ± 0.1 keV by linear interpolations.

[c] The subshells included into this table are numbered according to the Siegbahn notation. Contributions due to electron transfers from beyond the M shell have no significant effect on $(E_{KX})_{av}$, the energy of the peak displayed by NaI(Tl) detectors.

$$(E_{KXav})Ba \approx 38 \text{ keV} - 5 \text{ keV} \approx 33 \text{ keV}. \qquad (3.13)$$

The 33 keV KX ray peak is evident in the spectrum shown in Figure 3.4(d), so revealing the presence of barium in the X ray source.

3.9.3 Auger electrons and fluorescent yields

As noted earlier, the binding energies liberated following the transfer of atomic electrons are not only carried away as fluorescent X rays but also by atomic electrons as first identified by the French scientist P. Auger. These electrons are now known as Auger electrons. The ratio of the X ray to Auger electron intensities should be known if one works with either one or the other (NCRP, 1985, Appendix 1.2.8).

For each chemical element there is a well defined probability that, following a vacancy in any of the shells of its atoms, the vacancy will be filled with the emission of fluorescent X rays rather than Auger electrons. This probability is known as the total fluorescent yield, ω_t, obtained as the linear sum of its components contributed from each shell (ω_K, ω_L, . . .). For the present purpose only ω_K is of significant intensity (Figure 3.12). With ω_K known, the remaining K events, $1 - \omega_K$, are emitted as Auger electrons. Fluorescent yields are listed in nuclear spectroscopy tables, which are currently most conveniently accessed via the Internet (see Appendix 3).

For small atomic numbers, say $Z < 6$, the fluorescent yield ω_K is close to zero so that transitions occur effectively by Auger electron emissions only. The fluorescent yield increases steeply with increasing Z number, reaching about 88% at $Z = 56$ and approaching 100% at $Z > 80$ (Figure 3.12). Fluorescent yields have to be noted because K Auger electrons have some 20% lower energies than the KX rays from the same radionuclide, e.g. that of ^{55}Fe. Also, they have, of course, very different ranges in matter even when they have the same energy (Section 1.3.5). To display an Auger electron spectrum requires use of a silicon spectrometer, to be described in Section 5.5.2.

Chapter 4

Nuclear radiations from a user's perspective

4.1 The penetrating power of nuclear radiations

Chapter 3 dealt with selected properties of ionising radiations emitted by radionuclides with emphasis on the properties of X and γ rays and this emphasis will continue in the present chapter.

Gamma rays are useful for applications because of their high penetrating power (Section 1.3.5). Other applications utilise the short or very short ranges in matter of α or β particles, e.g. for the manufacture of paper or plastic materials to be discussed in Section 7.3.2. The large majority of industrial applications require radiations that are capable of transmitting information to a detector after travelling through a few centimetres of a metal or equivalent thicknesses of other materials, which is most effectively done by sufficiently energetic γ rays, about 0.5 MeV or higher.

Electrically uncharged neutrons also have high penetrating powers, although they are strongly absorbed by some elements (Section 1.3.6). Being equal in mass to protons, their interactions with matter are subject to different rules than that of massless γ rays (Table 1.2) Additional information about neutrons will be given in Section 5.4.4 in preparation for the numerous important applications to follow in Section 7.4.

It is again emphasised that radionuclides and the emitted radiations should only be used if one is sufficiently familiar with their characteristics (see below). Familiarity is also required with methods for the preparation of radioactive sources, with the success or failure of a project often depending on how well this has been done.

Table 4.1. *Summary of decay data for phosphorus-32, cobalt-60, caesium-137 and barium-137m. (NCRP 1985 Appendix A3. See also Figure 3.4.)*

Radionuclide and half life	Radiation	Energy (keV)	Intensity (%)
^{32}P 14.3 d	β^-	max. 1710 ave. 695	100.0
^{60}Co[a] 5.27 y	β^-	max. 318 ave. 96	99.9
	γ	1173.2 1332.5	99.9 100.0
^{137}Cs 30.0 y	$(\beta^-)_1$	max. 512 ave. 174	94.5
	$(\beta^-)_2$	max. 1173 ave. 415	5.5
^{137m}Ba 2.55 m	Auger L Auger K ce-K ce-L ce-M KX rays LX rays	3.7 26 624 656 661 33 4.5	7.6 0.8 8.0 1.5 0.5 7.3 1.0
	γ	661.7	90.1

[a] Two very low intensity β branches (<0.1%) have been omitted as well as very low intensity γ rays (<0.02%).

4.2 Radioactive sources

4.2.1 Radionuclides and their decay schemes

For the applications discussed in this book the commonly required decay data include the half life of the radionuclide and the types of emitted radiations with their energies and intensities – rarely much more than that. It is always helpful to list the decay data for all radionuclides expected to be required for an experiment in a table similar to Table 4.1. Caesium-137 may be required only for the 662 keV γ rays, but it is easy to make errors unless one is aware of the other emitted radiations shown in Figure 3.4(c).

There are numerous books available that list collections of nuclear decay data. As regards intensities and energies of X and γ rays, only collections obtained since the mid-1970s with high-resolution semiconducting detectors can still be used with confidence. Gamma ray data obtained before the 1970s with NaI(Tl) detectors missed numerous peaks (Figure 3.9(b), (c)), that can only be identified with high-resolution detectors to be discussed in Section 5.5.

The references listed at the end of this book include γ ray data from the 1970s onwards since many industrial laboratories still rely on earlier texts. The data produced by ICRP (1983), NCRP (1985, Appendix A3) and Debertin and Helmer (1988, Table 6.6) are all up to date.

The *Charts of Nuclides* list all stable and unstable nuclides identified at the time of publication, currently some 280 stable and over 2500 unstable nuclides. A small section from such a chart is shown in Figure 1.3 (CoN, 1998). Readers are also referred to *Nuclides and Isotopes* (1998), produced by the General Electric Company Nuclear Energy Operations. These charts have only space for a few of the principal items of decay data (Figure 1.3), but they also offer other useful data and provide easily accessible information on potentially available radionuclides.

Most recently published decay data were derived from the Evaluated Nuclear Data Structure File (ENDSF) maintained and continuously updated at the Brookhaven National Laboratory (BNL), PO Box 5000, Oak Ridge, Tennessee 37830, USA, available via http://www.nndc.bnl.gov. However, data needed for routine applications are, as a rule, just as easily obtained from the above mentioned printed collections as from a web site which is likely to include a large amount of data beyond what is required for routine purposes. Notwithstanding this, there are now many occasions when access to electronically stored data is necessary or desirable. General information on access to electronic data is given in Appendix 3, with Table A3.1 listing Internet addresses.

4.2.2 Source making and counting procedures

Laboratory equipment

Radionuclides are normally supplied and stored as aqueous solutions. It is then necessary to make dilutions or evaporate solvents and dispense aliquots by weight or by volume, to make radioactive sources which have the required activity, chemical composition and physical form and which satisfy radiation protection requirements (IAEA, 1996a, Appendix 4). Frequent requirements are for solvents of adequate chemical and radionuclidic purity, which are prerequisites for the preparation of thin radioactive sources. Sources that are adequate for penetrating γ rays are not sufficiently thin for β particles (see below), where there is always the low-energy section of the spectrum to be considered (Figures 3.6(a) to (d). Even thinner sources are required for α particles (Section 1.3.5).

Two instruments essential for source preparation are a precision balance to prepare dilutions and sources and a still for solvent purification. Currently offered laboratory balances have built-in standard weights to verify the calibration. Weighings at accuracies of 0.1% or better also require buoyancy corrections. To improve the quality of distilled water from a standard laboratory still, should this be needed, distillations should be made in the temperature range 50 to 75 °C. They are slow, but total solids in the distillate can be kept down to less than 50 μg per gram of solvent.

Aliquots can be dispensed using disposable polyethylene micropipettes available down to 0.5 μl and stated to be accurate to ±1%. The use of disposable equipment reduces risks from contamination when working with different solutions. In assessing the amount of activity to be purchased for some project, generous allowance should be made for errors during source preparations and losses due to source decay.

Procedures for making thin sources

If thin $2\pi\beta$ sources are required, subject to minimum backscatter and self-absorption, they should be prepared on 0.5 or 1 mg/cm^2 Mylar film glued onto thin light metal rings. The radioactive solution could be dispensed by volume using a micropipette. The solvent is dried to leave a deposit which should be made as uniformly thin as possible. This can often be achieved with the help of a drop of dilute wetting agent, e.g. insulin, or a seeding agent, e.g. Ludox ^{SM}, added to the radioactive aliquot prior to drying (Mann *et al.*, 1991, Section 3.3.3). To count the activity, the source is placed in a detector, e.g. a 2π proportional counter (half of the 4π counter shown in Figure 6.1(a)). Thin 4π sources are needed to obtain accurate activity measurements of α or β emitting radionuclides. Supports for 4π sources are made from a plastic material known as VYNS which can be prepared as thin as 10 to 20 μg/cm^2.

Source self-absorption when counting α or β particles can be avoided by liquid scintillation (LS) counting, a method which combines the radionuclide and the detection medium, an organic scintillant, as solutes into the same solution. Some aspects of this method will be described in Section 4.2.4.

Another method for making thin 2π sources, especially from metallic radionuclides, makes use of electroplating techniques. However, plating efficiencies may not be as reproducible as required and there could be problems with backscatter.

When samples of γ ray emitters are required in large numbers and in highly reproducible source–detector geometry, this is best done with the radionuclide in solution. These solutions are dispensed as small equal volumes into commercially available thin walled plastic vials when the countrate can be measured in a well-type NaI(Tl) detector in close to 4π geometry. Countrates should be no more than a few hundred counts per second (cps) to avoid corrections for accidental summing. Necessary corrections to countrates are discussed in Sections 4.5.3 and 4.5.4.

Computer controlled γ ray detection systems can identify small differences in the properties of large numbers of samples of solutions dispensed as described above. A suitable γ ray emitting radionuclide is selected to trace the property of interest, e.g. differences in the concentration of specific contaminants. The detector is shielded to keep background counts low while obtaining the required information in counting times of 100 s per sample or less. It is not practical, however, to use similar systems for routinely counting β particles or even low energy photons, say below 20 keV.

4.2.3 Sealed sources

Numerous applications require high-activity portable sources when the encapsulation must be leakproof or radiation protection authorities will not permit them to be used (IAEA, 1996a). When employing sealed sources, radionuclides have to be selected to ensure that they have the right properties for the task, such as the following.

- *The specific activity of the radionuclide (MBq/g)*: this may have to be maximised in order to minimise the volume of the source and also source self-absorption effects.
- *The types and energies of the emitted radiations*: a high-activity α particle source (alpha particles are all in the megaelectronvolt range) requires an encapsulation with very different properties from what is required for a high-activity, high-energy (>1 MeV) γ ray source (see below).
- *The half life of the radionuclide*: this should not be less than a few years because encapsulation is expensive and the cost of a source is always an important item.
- *Emitted radiations other than the radiations of interest*: such unwanted radiations could be unavoidable, but it is often possible to find a radionuclide that emits radiations other than the radiation of interest only at non-interfering energies and intensities. For a list of selected γ and β ray emitters which could be useful for this purpose, readers are referred to Tables 8.1 and 8.2 in this text.

Sealed sources are offered commercially, made with a large range of characteristics (see list of suppliers in Section 4.3.3). There are the ^{241}Am α particle sources ($T_{1/2} = 433$y) of activity upwards to 10^7 Bq/cm^2, made by sealing a 2 μm thick layer of americium metal between two about 2 μm thick films of a noble metal which transmit some 60 to 70% of incident α particles

(Charlton, 1986, Figure 12.1). If carefully used these sources are leakproof. They are in strong demand not only because of their long half life, but also because ionisations by α particles can eliminate electrostatic effects, e.g. in the air of precision balances, or act as smoke detectors while the γ rays following ^{241}Am decays are commonly of sufficiently low energy and intensity not to cause problems (Figure 3.1).

Usefully thin sources are also available for low-energy β particle emitters. On the other hand, there are securely sealed high-energy, high specific activity ^{60}Co γ ray sources ($\sim 10^{10}$ Bq/g) used in hospital teletherapy units to destroy deep-seated cancers. Such sources have to be adequately shielded to protect personnel, but they must permit a narrow beam of radiation (achieved in part by carefully designed collimators), to reach the patient with minimum contributions from scattered radiation (Section 4.4.4), to avoid damaging tissue and to realise high peak-to-total ratios for the 1.33 and 1.17 MeV γ rays, since the full-energy peaks are therapeutically most effective.

4.2.4 Liquid scintillation counting to minimise source self-absorption

The development of liquid scintillation (LS) counting during the 1950s and 1960s will be discussed in more detail in Section 5.4.2, but since this section deals with methods for minimising source self-absorption, a few words on LS counting are appropriate here.

LS counting made it possible to routinely employ the low-energy β particle emitters tritium (^{3}H, $E_{\beta max} = 18.6$ keV) and carbon-14, (^{14}C, $E_{\beta max} = 157$ keV) which are now widely used, e.g. for environmental surveillance (Sections 9.3 and 9.4). Quantitatively prepared solid sources are commonly unsatisfactory because, at these low energies they are subject to uncertain self-absorption effects.

Routine LS counting uses clear glass or clear plastic vials seen by two photo-multiplier (PM) tubes connected in coincidence (Figure 5.4(b)). The method employs the radionuclide and the scintillator dispersed or dissolved in the same organic solvent. The radionuclides are not only pure β particle emitters but also sources of α particles and of low-energy X or γ ray emitters. Source self-absorption effects are avoided since the radiations can excite scintillator molecules without having to expend more than the few electronvolts needed to excite the scintillants. The molecules de-excite with the emission of light quanta which, all being well, reach the PM tubes to trigger counts.

Problems arise because scintillant molecules do not only de-excite with the emission of light quanta but also by thermal effects, and light quanta are absorbed in the detector before they reach the photocathode of the PM tube

(Section 5.4.2). These effects cause the efficiencies of LS counting to fall well short of 100% notwithstanding the absence of source self-absorption. Losses in counting efficiency are larger the lower the average energy of the emitted radiation.

4.3 Gamma ray applications

4.3.1 The role of electronic instruments

When ionising radiations interact in a detector their energies are converted into signals which operate ratemeters or scalers or multichannel analysers. Signals are also stored for subsequent computer processing. This section offers a brief introduction to the principal functions of nuclear instrumentation normally used for routine applications of γ ray emitters detected with NaI(Tl) crystals. Figure 4.1 is a schematic diagram of a NaI(Tl) detector which illustrates the three γ ray interactions (γ_1, γ_2, γ_3) described in Section 3.4.3 and shown in Figure 3.7(a). Figure 4.2 shows a block diagram of the associated signal processing equipment.

The cathode ray oscilloscope (CRO) serves to monitor pulses at any of the terminals. The very short negative or positive going parts of the tail pulses shown in the figure are due to the rapid collection (within a fraction of a microsecond) of the very light and mobile photo-electrons emitted from the photocathode of the PM tube and multiplied in the strong electric field across its dynode chain. The very much heavier positive ions, the molecules which lost an electron, take some 10^4 times longer to be collected. This is illustrated by the long tails which should be very much longer than could be shown in the figure.

The preamplifier is built into the base of the PM tube adjacent to the output terminal of the detector. It serves to process and transmit the range of pulses (down to submillivolt heights) generated by ionising radiations in the detector and this without significant pulse degradation or deformation which would occur in long cables (exceeding about one metre) if pulses which were not pre-processed were allowed to reach the amplifier.

The preamplifier also reduces the length of the tail pulses (clipping), so reducing pulse pileup. Moreover, it provides an input for test pulses (Figure 4.2) from a precision pulser, the output of which serves to monitor the proper functioning of all components of the system and to determine corrections for random pulse summing (Section 4.5.3).

The amplifier is the core of the pulse processing equipment. Spectroscopy amplifiers provide highly stable and linear amplification, but the amplified

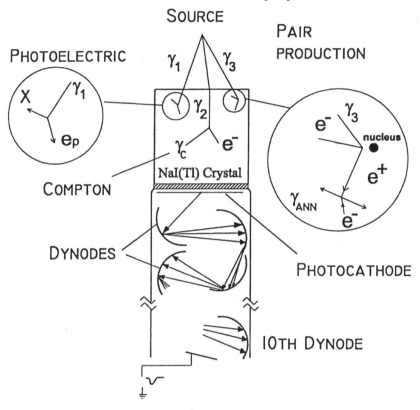

Figure 4.1. Gamma ray interactions in a NaI(Tl) detector shown with the associated electronics. The photoelectric effect (γ_1), Compton scatter (γ_2) and pair production (γ_3) are illustrated.

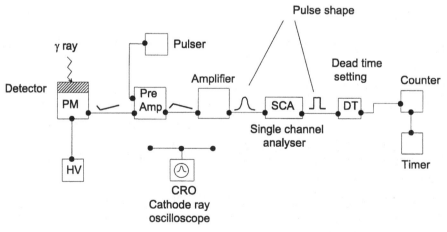

Figure 4.2. A block diagram of equipment used for the signal processing of pulses due to γ rays detected in a NaI(Tl) crystal (see text, Section 4.3.1 for explanations). The pulser and dead time gate (DT) are rarely required for routine work.

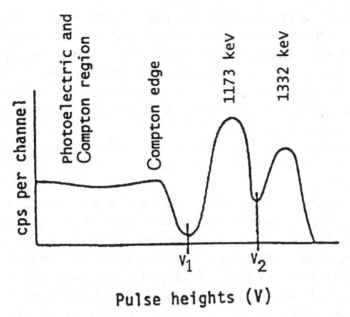

Pulse heights (V)

Figure 4.3.   A measurement of the countrate in the 1173 keV full energy peak of a spectrum due to ^{60}Co, using a single channel analyser (SCA), to define the window V_1 to V_2 (Mann *et al.*, 1980, Figure 7.13).

pulse at the output should not exceed 10 V to avoid amplifier overload. Another function of spectroscopy amplifiers is to shape the tail pulses from the preamplifier into Gaussian-type pulses (Figure 6.11). Tail pulses are not always readily accepted by the pulse analysers that follow the amplifier (Figure 4.2).

Electronic instrumentation gives rise to electronic noise pulses which have to be kept well below pulse heights due to signals generated by ionising radiations in the detector. Higher noise levels picked up from external sources can be greatly reduced by connecting all parts of the equipment to a common earth terminal with thick copper leads which should be as short as possible.

The single channel analyser (SCA) consists of a lower level discriminator (LLD) topped by an adjustable gate or window (W) or independently adjustable lower and upper level discriminators (LLD, ULD). In Figure 4.3, the LLD and ULD were set to obtain the countrate in the 1173 keV peak of a ^{60}Co spectrum (Section 4.4.3).

4.3.2 NIM bin and portable equipment

Signal processing instrumentation for signals from nuclear radiation detectors is normally used as NIM bin equipment (Nuclear Instrument Modules), an internationally adopted system. The frame for the modules fits up to twelve single width units and comes with or without power supply, with frame and modules fabricated in accordance with strictly defined specifications. Modules are available for each of the many different functions required for signal processing: power and high-voltage supplies, amplifiers, discriminators, SCAs, timers, scalers, pulsers and other more specialised instruments described in the catalogues of manufacturers.

With electronic equipment continuously being reduced in size, modules are now of reduced height ('slimline'), but existing frames and power supplies remain serviceable.

Numerous applications of nuclear radiations include investigations 'out of doors', requiring portable and rugged equipment which may have to be operated in a difficult environment. Field work employs robust, battery-powered stand-alone units incorporating the preamplifier, amplifier, LLD and scaler–timer and sometimes a single-channel analyser. Signal processing units powered from the mains are also used in industry and in school and undergraduate laboratories, as are many components of NIM equipment.

4.3.3 *Comments on instrumentation and its supply*

The range of nuclear instrumentation available for industrial and other purposes is much larger than is often realised. The following are just a few comments on the characteristics of radiation detectors and in particular scintillation detectors. The NaI(Tl) crystal detector is well known, efficient and stable, but it is only one scintillator among numerous others which have been developed over the past two decades. Several of the more recently designed detectors exceed NaI(Tl) in density and luminosity and are non-hygroscopic. Readers should obtain full information on these detectors from manufacturers, several of whom are listed below. Others can be identified from advertisements in the technical literature.

Thallium-activated NaI (0.2% Tl by weight, $\rho = 3.67$ g/cm^3) is probably still the most widely used material for γ ray detection. It is, however, strongly hygroscopic and so has to be protected from atmospheric moisture, as well as from mechanical damage. This has to be done while minimising any reduction in the efficiency with which the detector responds to incoming radiation (Section 4.4.1).

Silver-activated ZnS crystals serve to detect α particles, being too small to be sufficiently sensitive for γ radiations. Since the background is nearly 100% due to γ rays (Section 4.6.4), these small crystals can be used unshielded, assuming that there are no other radiations likely to cause interference with the signals of interest.

CsI(Tl) crystals are only slightly hygroscopic but, other things being equal, they have barely half the light output of NaI(Tl) crystals. Scintillators are also made from organic crystals dissolved in a suitable solvent, notably anthracene and trans-stilbene, which are useful β particle detectors of very short dead time (<0.1 μs) but of too low a density (1.1 to 1.3 g/cm³) for routine γ ray detection, though there are the scintillation detectors referred to earlier. Plastic scintillators loaded with a few per cent of ^{6}Li or ^{10}B atoms are useful detectors for neutrons.

Manufacturers developed models of NaI(Tl) detectors which can be used under water or can tolerate mechanical vibrations or shocks or temperatures up to 200 °C. Progress has been made with high-density scintillators, which are efficient detectors for high-energy γ rays and which have come into widespread use over the past few years. In particular, BGO, a compound of oxides of bismuth and germanium, is commonly used.

The high Z number of bismuth ($Z = 83$) and the high density of bismuth germinate crystals (7.4 g/cm³, twice the density of NaI) make this crystal a much more efficient detector of higher energy γ rays than is NaI(Tl). It is fully transparent to its own radiation, non-hygroscopic (requiring no airtight seal), and easily machined, but it is expensive and its light output is only about 15% of that obtained from NaI(Tl) crystals in similar conditions. BGO is widely employed for the tomographic imaging of 511 keV annihilation γ rays (Section 1.4.4). Other high-density phosphors are being developed to obtain a higher light output than is available from BGO crystals. High-resolution detectors will be discussed in Section 5.5.

Information about developments in instrument technology is regularly published by suppliers of such equipment in their highly informative catalogues. These firms advertise extensively in the scientific literature and many of them maintain service departments all around the globe. Here it is only possible to offer a very small selection without prejudice. For example:

- Canberra Industries Inc., and its subsidiary, Packard Instrument Company Inc., Meriden, CT 06450, USA.
- PerkinElmer Instruments, Oak Ridge, TN 37831-0895, USA (formerly EG&G Ortec Nuclear Instruments).

Other manufacturers specialise in radiation detectors and pressurised ionisation chambers.

- Bicron Scintillation Products, 12345 Kinsman Rd, Newbury, OH 44065, USA.
- NE Technology Ltd, Reading, Berkshire RG7 5PR, England.

- Centronic Ltd, New Addington, Croydon CR9 0BG, Surrey, England
- Nuclear Associates, PO Box 349, Carle Place, NY 11514-9895, USA
- Capintec Inc, Arrow Rd, Ramsay, NJ 07446, USA.

Radiochemical suppliers include:

- Amersham International plc, Little Chalfont, Buckinghamshire HP7 9NA, England
- Mallinkrodt International, 675 McDonnell Blvd, St Louis, MO 63134, USA
- Nordian International Inc, March Rd, Kanata, Ontario, Canada K2K 1X8.

4.4 Gamma ray counting with NaI(Tl) detectors

4.4.1 Further comments on NaI(Tl) detectors

NaI(Tl), an inorganic scintillation detector

When first employed, scintillation detectors were commonly organic crystals of low density, e.g. naphthalene, $\rho = 1.14$ g/cm^3, suitable mainly for the detection of β particles and lower energy X rays, up to 10 keV or even less. During the 1940s it was discovered that scintillations triggered in crystals of NaI could make them useful γ ray detectors because their density was adequately high ($\rho = 3.67$ g/cm^3). Also, the addition of 0.2% by weight of thallium ($Z = 81$) made even very large NaI crystals (Figure 4.4) fully transparent to the light quanta generated by γ rays throughout their volume. However, the crystals have to be kept free from moisture and there are other fabrication problems to be overcome.

Selected characteristics of integral assemblies

Routinely used NaI(Tl) detection systems are often known as integral line assemblies or just integral assemblies, combining the cylindrical detector, the PM tube and the preamplifier in a metallic shield. A frequently used design is to protect the cylindrical surface of the crystal (Figure 4.1) by a 2–3 mm thick aluminium sheet the inner surface of which is pre-treated to maximise its reflectivity. A 3 mm thick Pyrex or quartz window, selected to optimise the transmission and focussing of light pulses, connects the NaI crystal to the PM tube, the latter being protected by an outer Mumetal shield which extends around the preamplifier.

The entrance window is made as thin as practical to transmit photons with minimum attenuation. For laboratory applications the thinnest and rather fragile window is 0.25 mm beryllium, which transmits 5 keV X rays with close to 80% efficiency. Windows for detectors used in the field are made

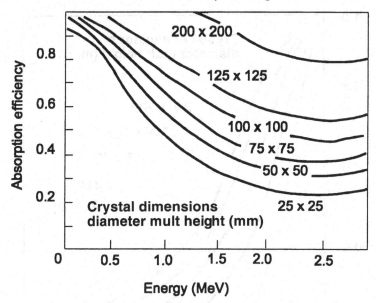

Figure 4.4.   Total detection efficiencies of NaI(Tl) detectors for collimated γ rays as functions of the dimensions of the crystals in millimetres (shown on the curves) and the energies of the γ rays (from catalogues of manufacturers).

from 0.25–0.5 mm Al, the former transmitting 20 keV photons with 80% efficiency. The corresponding efficiency of a 0.5 mm thick window is 60% (Hubbell 1982, see also Bicron catalogues). All assemblies are engineered to optimise light collection while minimising light losses.

On average it is necessary to absorb about 0.3 keV of γ ray energy in the NaI(Tl) crystals to produce a countable voltage pulse at the anode (Figure 4.1). The light output from NaI(Tl) crystals peaks at a wavelength near 430 nm (at the blue end of the optical spectrum), the wavelengths at which the PM tube is at its most sensitive (Neiler and Bell, 1966).

Total efficiencies and peak-to-total ratios

The efficiency with which NaI(Tl) detectors respond to γ rays increases with their volume, as shown in Figure 4.4. The efficient detection of high-energy radiations (>1 MeV) is favoured by larger detectors (75 mm × 75 mm or larger). Lower energy radiations (<200 keV) can be efficiently detected by 25 mm × 25 mm or smaller detectors with a compensating advantage that the smaller the crystal the lower the countrate due to the background.

To ensure as high a γ ray detection efficiency as practical for applications in the field, it is normal practice to include the entire pulse height spectrum (above a low LLD setting) in the count, a procedure known as integral

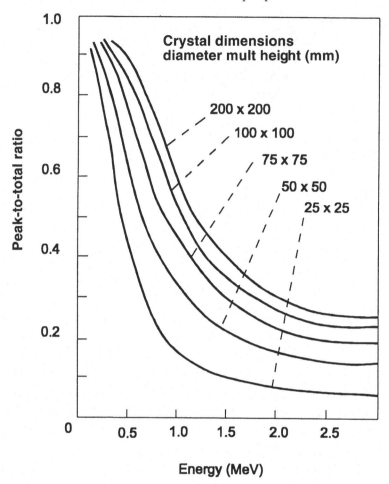

Figure 4.5. Peak-to-total ratios for NaI(Tl) detectors as functions of the dimensions of the detectors and energies of the γ rays. These ratios apply for narrow beam geometry (Figures 4.6(a)) but depend only weakly on the source–detector distance (Neiler and Bell, 1966).

counting (Section 4.4.2). Integral counting normally requires shielding of the detector, preferably behind 50 mm thick lead to reduce the background count which could exceed 400 cps in unshielded 50 mm × 50 mm detectors. Procedures to realise spectra with optimum peak-to-total ratios (Figure 4.5) will be referred to in Sections 4.4.3 and 4.4.4.

4.4.2 Integral counting

Gamma rays are emitted with a precisely defined energy (Section 3.5.3), but their detection results in a spectrum from near zero to the upper edge of the highest full energy peak (Figures 3.9(b), (c)), with many spectra due to multiple γ ray emitters and so including large numbers of peaks (Section 3.7.3). When a NaI(Tl) detector is required to detect the γ rays from a radionuclide with maximum efficiency, the amplifier gain and the discriminator levels should be adjusted to ensure that countrates include the entire pulse height spectrum generated in the detector (Section 3.5.3), assuming there is no overlap with other spectra.

The low-energy limit for photons to trigger a NaI(Tl) detector is fixed by the thickness of the entrance window, but low-energy pulses are also due to interactions within the detector material. They are consequences of Compton scatter and photoelectric effects, which account for the low-energy section of γ ray spectra, e.g. the ^{137}Cs–^{137m}Ba spectrum shown in Figure 3.4(d), together, in this case, with fluorescent X rays (Section 3.9).

Integral counting offers two useful advantages – a high counting efficiency and the use of the relatively simple electronics shown in Figure 4.2 which are available in NIM bins or as a compact, easily portable unit. Before applying a high voltage (HV) across the PM tube the correct polarity must be ensured. It is helpful, if not essential, to have a CRO available to monitor the output.

The operating voltage range (the plateau) for these detectors is normally at least 200 V, beginning between 800 and 1200 V. If it is known to begin near 800 V, the HV is set initially to 400 or 500 V and then increased, taking counts every 50 V. The countrate increases sharply to begin with, but as the operating voltage plateau is reached the rate of increase should be down to about 1% per 100 V, when all pulses due to γ rays exceed the LLD and are counted. The γ ray source used for the setting up procedure should be placed a few centimetres in front of the detector window and cause a countrate between 400 and 800 cps at the operating voltage.

The growth of pulses can be monitored using the CRO. On reaching the operating voltage (the plateau), the pulses should be of nearly identical height, allowing for random variations. Noise pulses should be preferably at least an order of magnitude smaller. The plateau ends at an HV value when noise pulses are sufficiently amplified to pass the lower discriminator, so causing a rapidly increasing countrate.

Having identified the position of the voltage plateau, the operating voltage should be selected within the plateau, say half way along, bearing in mind that pulse heights at the amplifier output must remain below 10 V. For example, to count a ^{60}Co spectrum (Figure 4.3) the pulse heights should not exceed 0.6 V per 100 keV of γ ray energy, since the upper edge of the 1332 keV peak would then be close to about 9.0 V, rather close to the 10 V limit.

If an upper-level discriminator is used, the cut-off energy should be high enough not to interfere with the spectrum. It is also important to bear in mind that pulse heights from a NaI(Tl) detector are only an approximately linear function of the γ ray energy (Section 6.4.4).

4.4.3 Peak counting

Peak counting is the term used when only the γ rays that generate pulses within one or more selected full energy peaks are counted.

For routine applications peak counting normally employs a NaI(Tl) detector and an SCA. Adding a CRO and a precision pulser is likely to lead to more satisfactory results (Section 4.4.2). If a semiconducting detector is used, a multichannel analyser (MCA) and computerised signal processing equipment are required to cope with the large amount of data. If at all possible, peak counting should be done in narrow beam geometry, since pulses due to scattered γ rays are lost from the peak. If scattered γ rays are unavoidable, the measurements should be taken in conditions where con-tributions from scattered γ rays can be reproduced as closely as possible (Section 4.4.4).

Assuming that the 1173 keV peak from a ^{60}Co source is to be counted (Figure 4.3), the amplifier is used to place the centre of this peak close to 5 V, so keeping the spectrum well below 10 V. Next, the SCA is set for a 0.10 V window (about 0.23keV). The window is moved across the peak, taking short counts every 0.2 V and is then set to the position where the short count was a maximum which is the centre of the peak. Next, the window is opened to encompass the range V_1 to V_2, when peak counting can be effected between these limits. For the spectrum shown in Figure 4.3 the backscatter peak is all but absent, indicating that pulses due to scattered γ rays had been minimised (see also Section 4.4.4).

Given the relatively low resolution of NaI(Tl) detectors, the positions of the low- and high-energy edges of full energy peaks (and therefore the positions of V_1 and V_2) are subject to uncertainties that could reach several per cent but could be greatly reduced when working with the well defined peaks due to high-resolution germanium detectors (Figure 3.9(b)). However, ^{60}Co peaks can often be counted without shielding, even in a laboratory, since at these high γ ray energies the background rate is almost negligible (Section 4.6.4).

4.4.4 Precautions to avoid errors due to Compton scatter

Integral and, in particular, peak counting yield the most accurate results if carried out in near scatter-free conditions, i.e. subject to narrow beam

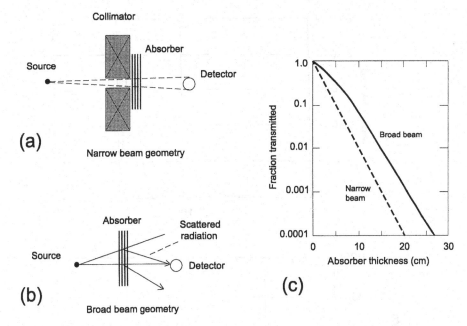

Figure 4.6. Gamma ray attenuation in narrow and broad beam geometry. (a) Narrow beam geometry. A collimator prevents γ rays scattered out of the beam by Compton interactions from reaching the detector. (b) Broad beam geometry when Compton scattered γ rays could be scattered back into the detector. (c) Gamma rays attenuated in iron in both narrow beam and broad beam geometry, illustrating build-up of the broad beam.

geometry (Section 3.4.5). However, there are many applications when γ rays have to pass through attenuating material before reaching the detector, and in such cases Compton scatter is unavoidable.

When γ rays are scattered in low Z number materials ($Z < 15$), significant Compton scatter begins at energies above about 50 keV (Figure 3.7(a)), when γ rays deflected from the beam can be scattered back into the detector so reducing the peak-to-total ratio.

If the intensities of γ rays transmitted in broad beam geometry (Figure 4.6(b)) are to be measured reproducibly via the countrate in the full energy peak, two points have to be noted. (1) Since Compton scatter reduces the peak-to-total ratio of the spectrum, it must be kept constant during a series of comparisons by ensuring a constant environment. (2) Results obtained using published attenuation coefficients, e.g. those shown in Figure 3.7(a), could overestimate the attenuation since they assume narrow beam geometry.

Build-up is defined as the ratio of the γ ray countrate following attenuation in broad beam geometry (Figure 4.6(b)) to the countrate that would have been observed for attenuation in narrow beam geometry (Figure 4.6(a)).

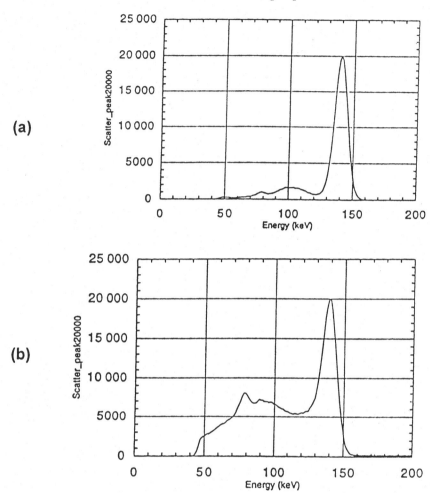

Figure 4.7. Pulse height spectra due to technetium-99m γ rays. Peak heights were normalised to a maximum of 20 000 counts per interval. (a) Spectrum due to a thin source in air and in narrow beam geometry, and (b) with the ^{99m}Tc dissolved in a water phantom when the peak-to-total ratio is greatly reduced by Compton scatter (courtesy of S. Eberl, Royal Prince Alfred Hospital, Sydney, Australia).

Build-up factors for various materials are available in the literature (e.g. Shleien, 1992). These factors increase with the thickness of absorbing material as shown in Figure 4.6(c) for iron. Every gamma ray scattered prior to detection reduces the peak-to-total ratio in the spectrum as shown in Figures 4.7(a) and (b).

Figure 3.15 demonstrates the effect of a high rate of Compton scatter on the appearance of spectra. The Compton region of the spectrum, i.e. the

section between the backscatter peak and the Compton edge (Figure 3.6(f))
is, in this case, a much more prominent feature of the spectrum than the 411.8
keV full energy peak. In a scatter-free environment the peak-to-total ratio at
412 keV is close to 0.6 (Figure 4.5) whereas it is only about 0.12 in Figure
3.15.

4.5 Corrections and precautions, part 1

4.5.1 A summary

Results of measurements of ionising radiations are nearly always incomplete and could be misleading until corrected for a number of effects,
notably radionuclide decay and unwanted or lost counts. Here it is only
practical to deal with the most commonly occurring corrections which
include:

1. dead time corrections
2. corrections for pulse pile-up, random and coincident summing
3. decay corrections
4. corrections to account for unwanted radiations.

Corrections 1 to 3 will be dealt with in this section and the fourth group in
is covered in Section 4.6. The order does not imply an order of importance.

4.5.2 Dead time corrections

If nuclear radiations trigger a pulse counter, they initiate a train of events
which has to go to completion before the counter can respond to another
ionising event (Section 5.3.3). The time during which it cannot process
incoming radiations is known as its dead time. Ionisation chambers operating
in the saturation region (Section 5.2.1) are not subject to dead time as defined
here. Semiconductor detectors are not subject to dead time either, but they
are subject to pile-up and summing effects.

Ionisation chambers are not subject to dead time because the electric field
across the chamber continues to collect ions formed in the gas regardless of
what went on before. Errors can occur if the field is not strong enough to
collect all the ions before they recombine, so causing the extent of the
ionisation to be underestimated.

The dead times of GM detectors range from about 50 to over 300 μs
(Section 5.3.2). The dead times of proportional counters similar to those
shown in Figures 5.2 or 6.1 are likely to be in the range 2 to 4 μs (Section

5.3.3). The dead time of NaI(Tl) crystals is only about 1 μs while that of organic scintillators (Section 4.2.4) can be shorter than 0.1 μs.

If a radionuclide is available which has a well known and sufficiently short half life, one can measure the effective dead time for a detector as follows. Beginning with a high countrate which is subject to a large dead time correction, one takes periodic counts down to countrates where the dead time correction is small enough to be neglected. Using the known decay rate of the radionuclide, one can calculate the 'true' countrate for each measured result and obtain an average value for the dead time τ using Eq. (4.2).

An accurately known dead time can be important when countrates (>500 cps) have to be accurate within about ±2%. If so, equipment should be available to make the dead time adjustable and closely reproducible. This can be done using a commercially available dead time gate (marked DT in Figure 4.2). For other methods of measuring dead times readers are referred to NCRP (1985, Section 2.7).

As a rule, dead times can be assumed to be non-extending, being unaffected by ionising radiations entering the detector while it is dead. Corrections for nonextending dead time losses can be calculated from the simple expression

$$N = N'/(1 - N'\tau),   \tag{4.1}$$

where N cps is the corrected countrate, N' cps the observed countrate and τ is the dead time in seconds. If both N and N' are known τ can be calculated by rearranging Eq. (4.1) to obtain:

$$\tau = (N - N')/NN'   \tag{4.2}$$

Dead time corrections should always be applied before other corrections are applied.

4.5.3 *Pulse pile-up, random and coincidence summing*

Randomly occurring effects

Pile-up occurs in the amplifier. Unlike pulse counters, amplifiers remain open to incoming signals regardless of what went before, but their response is subject to a resolving time. Two pulses arriving within the resolving time are processed together leading to the pile-up and/or summing effects respectively shown in Figures 4.8(a) and (b). A resolving time close to 1 μs is short enough to avoid difficulties due to pile-up until countrates exceed 2000 to 3000 cps, depending on circumstances.

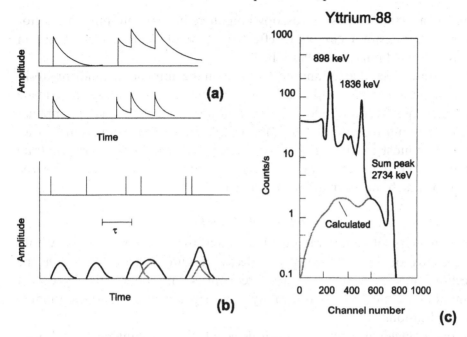

Figure 4.8. (a) Random pulse pile-up before and after pulse shaping by the amplifier. (b) Pulse superposition of two pulses to produce a sum pulse. (c) Pulse summing in a NaI(Tl) detector for γ rays from yttrium-88. The sum spectrum below about 2 MeV was obtained by calculation, allowing for the source–detector geometry.

Pulse distortions due to pile-up distort the spectrum while random pulse summing can cause counts to be lost as two pulses combine into a single pulse. If possible, these effects should be avoided by reducing countrates, since they are difficult to correct for. An example of pulse summing is shown in Figure 4.8(b) and Figure 4.8(c) shows a sum peak obtained with a ^{88}Y source and also the continuation of the summing region to lower energies (this section of the spectrum has to be calculated).

The spectrum in Figure 4.8(c) was taken in conditions where there is both random and coincidence summing. Random summing could have been minimised by reducing the countrate, e.g. by increasing the source–detector distance. Although coincidence summing can in this case be reduced in the same way, it is normally subject to other rules (see below).

Effects due to random summing are also reduced or avoided by reducing the amplifier resolving time. A relatively long resolving time, say 4 μs, improves the energy resolution of the detector, but it leads to a higher probability for pile-up. If high countrates are more important for an application than high-energy resolution, the resolving time should be as short as practical. High-class spectroscopy amplifiers incorporate pile-up rejector and base line restorer circuits to effect the type of improvement shown in Figure 4.8(a) (Mann *et al.*, 1980, Ch. 7).

Countrate losses due to random summing in the amplifier are indistinguishable from losses due to dead time in pulse counters. Corrections can be made using a 50 cps precision pulser. The pulses are injected into the preamplifier (Figure 4.2) and the output rate is measured using an SCA and scaler or an MCA. For most routine applications, the percentage loss in the countrate obtained from the spectrum can be taken as equal to that observed for the pulser for which the 'normal' rate is known.

Coincidence summing

The daughters of many radionuclides emit two or more γ rays within picoseconds or less following the decay (Figure 3.4(b)). Although the directions in which these γ rays are emitted are often correlated, the assumption of randomness will rarely cause significant errors during industrial applications.

If two coincident γ rays are both detected, the resultant pulses will sum since they arrive at the detector well within its resolving time. While random summing is a function of the rate at which photons reach the detector, coincidence summing is a consequence of the decay scheme of the radionuclide and of the detection efficiency.

The efficiency of radiation detectors is the higher the larger their surface and the smaller the source–detector distance. If ^{60}Co radiations are detected in a large well-type NaI(Tl) crystal, the countrate, which in this case would be strongly affected by coincidence summing, could be lower than 60% of the γ ray emission rate. By contrast, if the source–detector distance exceeds about 30 cm, the detection of more than one of the coincident γ rays is highly improbable. For a 50×50 mm detector the probability to detect two coincident γ rays is sufficiently small to be ignored for many purposes, even if the source–detector distance is only 15 cm. To avoid the need for what are likely to be uncertain corrections, counting should be carried out in conditions when coincidence summing is negligible.

Yttrium-88 decays are followed by two coincident γ rays, as are ^{60}Co decays. The conditions shown in Figure 4.8(c) are clearly such that the γ rays are subject to coincidence summing as well as the random summing mentioned earlier. A comprehensive discussion of summing effects is given by Debertin and Helmer (1988, Sections 4.5 and 4.6); they deal with semiconductor detectors, but much is applicable to work with NaI(Tl) detectors.

4.5.4 Decay corrections

Results of radioactivity measurements must refer to a reference time T_r, which has to be stated the more accurately the shorter the half life of the radionuclide. If T_r and $T_{1/2}$ are known, a result can always be corrected to tell operators what it would have been at the time of interest, say T_t. Such corrections are most easily made if T_r, T_t and $T_{1/2}$ are expressed as fractions of the year (365.25 d) rather than in the conventional way.

A measured countrate N_t is corrected to what it was or will be at some other time T_t, using equations (1.6) and (1.7) (Section 1.6.2) when it can be shown that N_t, the countrate at time T_t is related to N_r, the countrate at the reference time T_r by:

$$N_t = N_r \exp -(0.693\, \Delta t / T_{1/2}) \tag{4.3}$$

where $\Delta t = T_r - T_t$. If T_t precedes T_r, Δt becomes negative when the exponent becomes positive. The uncertainty in this correction is the greater the greater the ratio $(T_t - T_r)\,/\,(T_{1/2})$. T_r could be chosen at any convenient time but preferably at T_{st}, the time when counting starts, or at the mean of the counting time, especially so if the half life of the radionuclide is not well known.

4.6 Corrections and precautions, part 2

4.6.1 Unwanted radiations, a summary

Ionising radiations emitted by the radionuclide of interest never have the field to themselves. They are invariably accompanied by radiations from other sources, here called unwanted radiations which fall into one or more of three groups:

radiations from radioactive daughters or parents (including isomers)
radionuclidic impurities
background radiations due to a variety of sources.

4.6.2 Radioactive parents and daughters

When using a radioactive parent, its radioactive daughter could be responsible for unwanted radiations. To what extent this is so depends on the ratio between their half lives and this should always be verified.

For cases (b) and (c) in Section 1.6.1 and Figure 1.7, the parent cannot be used free from radiations due to the radioactive daughter. The daughter could be eluted and used on its own, but regrowth into the parent is too fast to permit the latter to be used free of daughter activity. If both parent and daughter are present, one should make measurements only when the two activities are in equilibrium (if possible), to help with avoiding errors when making decay corrections.

Parent activities are readily used if the half life of a recently eluted daughter is long enough to keep its activity negligibly small. This is so for two frequently used radionuclides, technetium-99m ($T_{1/2} = 6.0$ h) and americium-241 ($T_{1/2} = 432$ y). The daughters are the β^- particle emitting technetium-99 ($T_{1/2} = 2.14 \times 10^5$ years), and the α particle emitting neptunium-237 ($T_{1/2} = 2.14 \times 10^6$ years). Their percentage contribution to the decay rate of the parents is sufficiently small to be disregarded.

4.6.3 Radionuclidic impurities

The magnitudes of the impurities dealt with in this section are assumed to exceed the uncertainty estimate applying to the measurement (see Section 6.5.2), but not be high enough to be easily identifiable.

Measurements employing radioactive sources could be affected by chemical (non-radioactive) impurities, or by radionuclidic (radioactive) impurities when the radionuclide of interest is contaminated by an unwanted radionuclide. Chemical impurities could affect radioactivity measurements, e.g. if they increase source self-absorption or affect the chemical stability of source material. Radionuclidic impurities are likely to be the more difficult to detect, the more similar their half lives are to that of the principal radionuclide.

Gamma or X ray emitting impurities introduced accidentally when working with γ ray emitters can be identified from their energies when there is access to a γ ray spectrometer, preferably with a semiconducting detector. For many radionuclidic impurities detection is only possible by monitoring the half life of the principal radionuclide.

If a radionuclidic impurity is due to a pure β particle emitter with a high enough activity to cause errors during measurements, it is likely to affect measurements of the half life of the principal radionuclide. This could be detected by monitoring its half life as long as practical using the method which was employed for the application. To what extent such tests could be successful depends on the respective lengths of the two half lives and the difference between the β particle energies. If results of half life measurements give no indications of an impurity, radionuclidic impurities are unlikely to be significant. However, tests to verify radionuclidic purity should be continued for as long as a radionuclide is employed.

Reputable suppliers provide documentation stating all radionuclidic impurities that were detected but could not be eliminated. Users then must look up the decay scheme data of any residual impurities (Section 4.2.1), consider the characteristics of their detection equipment and estimate what has to be done to avoid errors.

4.6.4   The gamma ray background

Table 2.5 lists natural and man-made background radiations (background for short), but knowledge of how the background affects the results of activity or countrate measurements is also important. In addition there could be unwanted counts due to nearby radioactive sources. In practice, it is γ rays that are responsible for nearly all background counts. Charged particles from external sources have short ranges in matter and will only penetrate into the detector if emitted 'close by', but their possible presence should not be overlooked.

The background due to γ rays recorded in a built environment is commonly relatively high. An unshielded 50×50 mm NaI(Tl) detector operated for integral counting in an industrial or hospital laboratory rarely records less than 200 cps, with rates of over 400 cps quite possible (see below).

A large fraction of the photons that account for the background are emitted from naturally radioactive materials at energies up to 2.6 MeV (Tables 2.5 and 4.2). The concentration of naturally occurring radionuclides (Fig. 1.5), is, on average around 10 parts per million in the soil and in stones and minerals used for buildings. Radiations escaping from these materials are nearly always multiple Compton scattered by walls, floors and ceilings prior to reaching a detector. As a result the initially energetic γ rays are reduced in energy, on average to around 100 keV as is evident in the typical background spectrum shown in Figure 4.9, obtained with a 50×50 mm NaI(Tl) detector near the centre of a laboratory. However, the loss of γ rays energy is compensated by build-up due mainly to Compton scatter (Section 4.4.4). The countrates would have been higher near the walls.

Detectors used for γ ray applications and especially for γ ray spectrometry should be shielded by 50 mm thick lead bricks shaped to prevent 'shine' through gaps between the bricks. Similar structures serve to minimise radiation escaping from stores of radionuclides. Adequate shielding should also be used around liquid scintillation detectors since liquids are also efficient detectors of background γ rays.

Table 4.2. *Gamma ray energies and intensities emitted by uranium-238,*
thorium-232 and their radioactive daughters.[a]

[238]U[b]			[232]Th[b]		
Daughter nuclide	**Energy[c] (keV)**	**Intensity (%)**	**Daughter nuclide**	**Energy[c] (keV)**	**Intensity (%)**
[234]Th	63	3.8	[228]Ac	338	11
[226]Ra	186	3.5	[224]Ra	241	4.0
[214]Po	295	18	[212]Pb	87	7.9
	352	35		239	44
[214]Bi	609	45	[212]Bi	727	7.0
	1120	15	[208]Tl	510	23
	1238	5.8		583	85
	1378	3.9		2614	100
	1765	15			
	2204	5.0			

[a] The data in this table are based on NCRP, 1985, Appendix A3.
[b] Only those daughter nuclides are listed which emit γ rays at a rate exceeding 3.0% of the decay rate. The [235]U chain is disregarded, being of too low intensity (Section 1.4.1).
[c] Energies and intensities exceeding 10% have been rounded to the nearest integer.

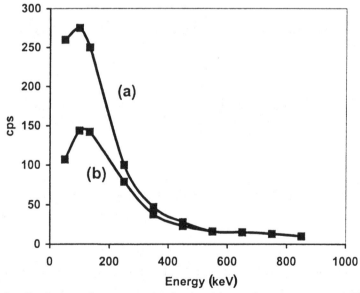

Figure 4.9. Background spectra obtained with a 50×50 mm NaI(Tl) detector. Curve (a) in a laboratory, with the detector unshielded. The peak near 100 keV is due to multiple Compton scattered γ rays. Curve (b) was obtained after covering the detector with 5 mm steel; the background rate is reduced by nearly a factor of two.

With background γ rays concentrated at energies around 100 keV (Figure 4.9), covering a γ ray detector with just 5 mm steel or its equivalent (when it remains easily portable) will cause a useful reduction relative to the unshielded background rate. When working in open fields and especially over water surfaces, the background rate could be two or more orders of magnitude below what is observed in the average laboratory. However, the background is relatively high near stone walls and especially so near deposits of uranium, thorium or mineral sand in the ground, or near megavoltage installations. Procedures to minimise background rates during low-level counting using anti-coincidence methods were described in detail by Watt and Ramsden (1964).

4.6.5 The alpha and beta particle background

Portable GM monitors used routinely as β particle detectors owe much of their enduring popularity to their low background rate, rarely higher than 5 cps even without shielding. This is due mainly to their low γ ray efficiency (Section 5.3.2) and the fact that background radiations are mainly γ rays (Section 4.6.4). For the same reasons the background rate is similarly low when working with proportional counters (Section 5.3.3).

Counts due to β particles or other electrons could be contributed by radionuclidic impurities in the material of the detector or by cosmic ray interactions and also by naturally occurring carbon-14 as $^{14}CO_2$, the supply of which is kept up by cosmic ray interactions in the upper atmosphere. Other β particle sources adding to the background are the long-lived tritium and caesium-137 left in the atmosphere from nuclear testing during the 1950s and 1960s (Section 9.4.2). Shielding β particle detectors by lead bricks is likely to reduce the background rate to a fraction of a count per second, which helps to keep it sufficiently reproducible, e.g. during comparisons of β particle countrates in proportional counters.

Proportional counters set up for α particle counting are operated at a lower polarising voltage than is required for the detection of β particles (Figure 4.10). The unshielded background rate during α particle counting is then unlikely to exceed 10 to 20 counts per hour, a small fraction of the background rate when counting β particles because electrons that enter a counter set up for α particle counting (e.g. those dislodged by γ rays) are 'ignored'.

The few background counts in α particle detectors are principally due to impurities which are naturally occurring α or β particle emitters, or to cosmic radiations interacting with atoms of the detector material, with counts

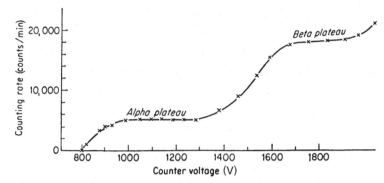

Figure 4.10. The counting of α and β particles in a windowless proportional counter at normal pressure. The α particle plateau begins just above 1000 V; the plateau for β particles begins near 1700 V.

resulting on the rare occasions when charged particles are emitted near enough to the inner surface of the counting chamber to permit them to reach the chamber and be counted.

Chapter 5

Ionising radiation detectors

5.1 Radiation detectors, a summary

Ionising radiations as normally used cannot be observed by the unaided human senses. They were discovered thanks to photographic emulsions which remain important detectors to this day, particularly in the biological sciences. There are many designs of detectors developed to respond to specific characteristics of the different types of ionising radiations discussed in earlier chapters and notably in Section 4.3.3.

Sections 5.2 and 5.3 will deal with ionisation detectors used for β and γ radiations, Section 5.4 will deal with liquid scintillation (LS) counting, adding to the material presented in Section 4.2.4. This will be followed by a description of microcalorimetry, an efficient detection method when working with high activities of many β and γ ray emitters. Section 5.4 will also cover the detection of neutrons at low and high intensities. Section 5.5 will deal with semiconductor detectors, which have long been indispensable for photon as well as for charged particle spectrometry.

Readers who seek details about specialist applications in any branch of radioactivity measurements are referred to the work of Knoll (1989), who deals with the scientific principles underlying these procedures over a broader range than other currently available textbooks.

5.2 Characteristics of ionisation detectors

5.2.1 Saturation currents and gas multiplication

A long established method for measuring the intensity of nuclear radiations uses their ability to ionise air or other gases such as argon or methane. Instruments that use an electric field to collect the ions and serve to measure

123

the resultant ionisation current are known as ionisation chambers, but ionisation detectors can also be operated as pulse counters.

When employed to detect γ rays, ionisation chambers should preferably be of several litres capacity. They will then have a satisfactorily high γ ray efficiency, especially when sealed and pressurised. Pulse counters are commonly small and serve principally for the detection of α and β particles and also of neutrons.

Pressurised ionisation chambers measure ionisation currents due to activities of the order of 10^{12} gammas per second and higher with reproducibilities within 0.10%. The stability of their response is due to their simple construction and operating characteristics (Section 5.2.2). They are useful γ ray detectors even though they are insufficiently sensitive for measuring ionisation currents due to low γ ray activities ($<10^3$ gamma rays per second) with the lower limit depending on the average energy and intensity of the detected photons.

Two types of radiation detectors which operate in the pulse mode are the proportional counter and the Geiger–Müller (GM) counter. A pulse counter needs only three or four ion pairs due to a charged particle which entered its sensitive volume in order to register a count with near 100% probability. This is so because detection is effected using ion multiplication by collision which is also responsible for the relatively long dead time of GM counters, setting a limit to attainable countrates (Section 4.5.2).

5.2.2 Three saturation chambers

Figure 5.1(a) to (c) show schematic diagrams of three types of ionisation chambers used in the saturation region for X or γ ray detection. They can be used open to the atmosphere because they can tolerate electro-negative gases such as oxygen which must be excluded from pulse counters (Section 5.3.3). The dimensions and other characteristics of commercially supplied chambers will be described in Section 6.3.2.

At low polarising voltages most ions formed by ionising radiations recombine before they can be collected. It is only when the voltage reaches the saturation level that it collects all ions to form the saturation current, which then remains constant even though the voltage is increased by another 200 to 300 V as shown in Figure 5.1(d). It is normal practice to operate saturation chambers at a voltage near the middle of the plateau. The voltage required to reach saturation depends slightly on the activity of the source; this should be borne in mind when working with both high- and low-activity sources.

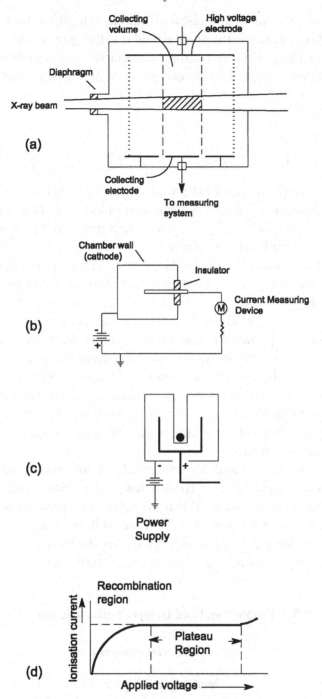

Figure 5.1. (a) to (c) Three designs of ionisation chambers operated in the saturation region. (d) A plot of ionisation current versus applied voltage showing the recombination and saturation regions. The plateau could extend over 300 V or more.

Assuming one measures a long-lived radionuclide, the saturation current remains constant as long as the applied field is sufficiently strong to collect all ions formed in the gas but not strong enough to accelerate ions to speeds where they have enough kinetic energy to ionise gas molecules. If this happens the ionisation current is increased beyond the saturation level so causing errors.

5.2.3 Parallel plate and cylindrical chambers

Parallel plate chambers open to the atmosphere are now only used for a few specialist purposes, e.g. for the measurement of exposure (Figure 5.1(a) and Section 2.4.2). The preferred type for the detection and measurement of X and γ rays is the sealed, cylindrical chamber where a relatively small diameter, hollow cylindrical anode which holds the radioactive source is surrounded co-axially by the cathode. In routinely used chambers, the anode (Figure 5.1(c)) is made of between 0.3 and 0.5 mm steel tubing which transmits photons down to about 15 keV with sufficiently small attenuation.

Sealed chambers permit the use of gases other than air and at above atmospheric pressure, so increasing the sensitivity of the measurements (Section 6.3). Also, the cylindrical geometry of the electrodes ensures that the electric field strength increases towards the smaller diameter anode (Eq.(5.1)). This arrangement results in a high efficiency for these chambers, as does the near 4π geometry. Both of these factors also ensure a high reproducibility of the results of measurements.

Ionisation currents caused by radionuclides in pressurised ionisation chambers cover a large range – from a few milliamperes measurable with routinely used current meters down to levels comparable with leakage currents which can be below 10^{-14} A in well maintained high-quality instruments (Section 6.3.4). Leakage currents set the limit to the accuracy of ionisation current measurements due to the radioactive source.

5.3 Proportional and Geiger–Müller counters

5.3.1 Thin wire counters

Operating principles

Figure 5.2(a) shows an ionisation chamber of the design used for pulse counters. The anode is a thin metal wire, commonly 0.02 to 0.1 mm diameter tungsten. Such thin wires have high electric fields at their surface even for

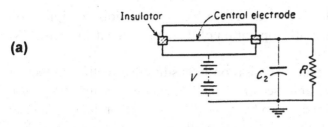

(a)

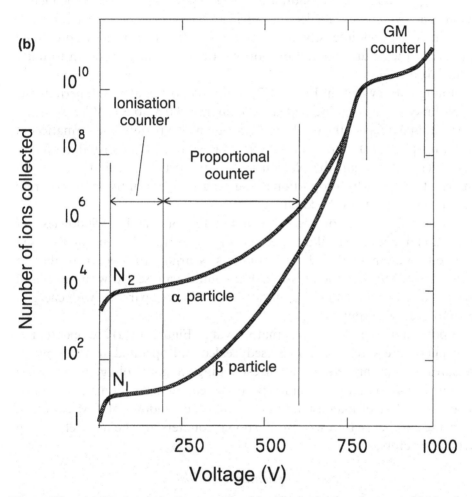

(b)

Figure 5.2. (a) A schematic diagram of a thin wire ionisation detector. (b) Numbers of ions collected per unit time in a thin wire counter as a function of the applied voltage when the primary ions are generated by α and β particles in the ratio 10^3 to 1. The three operating regions are indicated, though the plateau in the proportional region (Section 5.3.3) cannot be shown here.

relatively moderate applied voltages. These fields trigger ion multiplication by collision and characterise the operation of these instruments, usually GM and proportional counters.

Figure 5.2(b) shows a plot of ionisation intensities due to the number of ions collected (note logarithmic scale) in these chambers (per unit time) as a function of the applied voltage. Following the flat saturation region labelled ionisation counter, there is the onset of ion multiplication by collision leading to the proportional region followed by the steep increase towards the Geiger region. The two curves labelled β particle and α particle are here assumed to differ in ionisation intensity by a factor of 10^3. This ratio is preserved throughout the saturation and proportional regions but goes down to unity in the GM region.

Although not evident in Figure 5.2(b), the proportional region permits the establishment of a counting plateau, commonly 200 V to 300 V long, with a slope within 0.5% per 100 V. The Geiger region also permits the formation of a counting plateau (Figure 5.3(c)) though it is shorter (<200 V) and less flat (2 to 4% per 100 V) than the plateau in the proportional region. This is an important reason why proportional counting is significantly more accurate and reproducible than GM counting.

A typical GM detector is shown in Figure 5.3(a) and a windowless 4π proportional counter is illustrated in Figure 6.1(a). The normally used detection medium in sealed GM counters is argon–10% ethyl alcohol at about 10% of atmospheric pressure. The counters are sealed with metallised Mylar windows made thin enough to transmit charged particles with energies down to a few electronvolts.

Proportional counters such as that shown in Figure 6.1(a) are operated at atmospheric pressure, but can be used, sealed and operated at much higher pressures though this requires a correspondingly higher voltage to reach the plateau. The 4π sources used in proportional counters (Section 5.3.3), can be made two orders of magnitude thinner than the windows of GM counters, this being another reason why proportional counters are the more efficient β particle detectors.

Ion multiplication by collision

This process is responsible for the operation of both GM and proportional counters. It is only in the very small volume, very close to the surface of the wire anode that the electric field is strong enough to cause ion multiplication by collision, so making the response of these counters effectively independent of the region where the ions were created, which is an essential requirement.

The electric field strength E_r at radius r mm from the axis (i.e. the centre

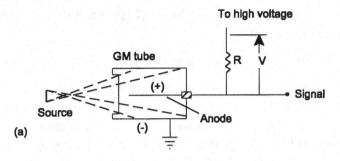

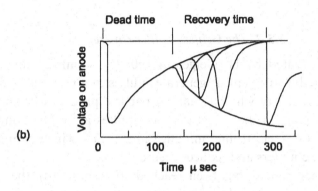

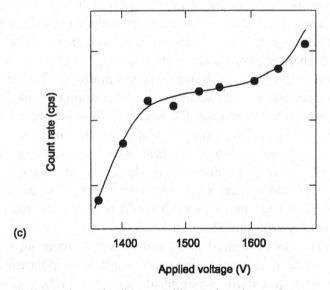

Figure 5.3.   Three schematic diagrams illustrating the performance of GM detectors. (a) The detector and its signal processing circuit. (b) The electric field at the surface of the anode immediately following a count. (c) A GM plateau for a 5 cm diameter tube.

line of the thin wire) of a cylindrical detector (Figure 5.2(a)) and subject to an applied voltage V, is given by:

$$E_r = V/[r \ln (b/a)], \tag{5.1}$$

where b and a are the diameters of the cathode and anode respectively. Assuming $V = 1000$ V, $a = 0.02$ mm and $b = 25$ mm, the field strength at 0.011 mm from the axis of the anode wire (0.001 mm above its surface) would be $1000/[0.011 \ln (25/0.02)] \approx 12\,750$ V/mm. At 0.2 mm from the axis of the anode, E_r is equal to $1000/[0.21 \ln (25/0.02)] \approx 670$ V/mm, or only about 5% of the first value.

5.3.2 The Geiger–Müller counter

The GM tube (invented by H. Geiger in 1908) is a simple and relatively cheaply constructed instrument. It generates pulses of a volt and higher with close to 100% probability when a charged particle produces no more than three or four ion pairs in the detector. These pulses are high enough to be detected without prior electronic amplification, a fact which simplifies the operation of these counters and reduces their cost.

When the Geiger plateau has been reached (Figure 5.2(b)), the tube not only produces readily detectable output pulses, but does so regardless of the number of ions responsible for triggering the counter. This is the reason why GM counters cannot be used to measure particle energies. In contrast, proportional counters can serve to measure β particle energies which can be the higher, the higher the gas pressure in the counter.

When the GM detector has been triggered, ion multiplication by collisions leads to intense ionisations and excitations in the counting gas at the surface of the thin anode wire, so reducing the electric field along the wire to near zero (Figure 5.3(b)). The situation is controlled by the ethyl alcohol in the counting gas, acting as a quenching agent which absorbs the very high concentration of ions and photons along the anode. Efficient quenching permits the electric field in small tubes (~10 mm diameter) to recover within 100 μs. Larger tubes (>25 mm diameter) could require over 200 μs with a correspondingly long dead time (Section 4.5.2).

GM detectors can be purchased with separate signal processing equipment including a dead time gate, so allowing users to adjust the polarising voltage and dead time as required for an application (Figure 4.2), e.g. in undergraduate laboratories. Most GM detectors are sold as radiation monitors, with just an on–off switch and a range switch for the scaler and a control button to verify the voltage setting at some value between 300 and 800 V,

Table 5.1. *First ionisation potential (E_i) and the average energy needed to create ion pairs (W) in selected gases.*

Gas	E_i (eV)	W^a (eV)
Argon	15.7	26.4
Carbon dioxide	14.4	34.4
Hydrogen	15.9	36.3
Methane	15.2	27.0
Nitrogen	16.7	34.7
Oxygen[b]	12.8	31.0

[a] These values apply to both α and β particles within about $\pm 5\%$.
[b] Oxygen molecules tend to attract atomic electrons with a small but significant probability. If so, the negative ions are some 3×10^4 heavier than free electrons. This probability is negligibly small for the other gases.

depending mainly on the counter geometry. Users can rely on modern, stable electronics which ensure that equipment malfunction is a rare occurrence.

5.3.3   The proportional counter

The proportional region

Proportional counters are operated at an applied voltage where the pulse heights at the output of the detector remain proportional to the number of primary ions generated in the detector by charged particles (Figure 5.2(b)). If the number of ions in the output pulse is N' when the number of primary ions is N, then:

$$N' = M \times N \qquad (5.2)$$

where M is a constant.

Being generated at a much lower gas gain than applies to GM counters, the output pulses from proportional counters are too small to be counted without prior electronic amplification. Also, while saturation ionisation chambers can be operated with atmospheric air as the counting gas, this is not so for pulse counters. They rely on the rapid collection of the negative ions at the anode, which is readily achieved with very light electrons. If the counting gas contains oxygen, its molecules attract electrons, so increasing the mass of the negative ions by factors of order 10^4 with a corresponding increase in their collection time which should be kept well below 1 µs. Times longer than 1 µs are unacceptable for satisfactory counter performance (see Table 5.1, Note b).

Other useful characteristics of these detectors are now noted.

Unsealed 4π windowless proportional counting

This method is useful when accurate countrates for large numbers of sources are requred (Figure 6.1(a)). The counting gas, e.g. argon–10% methane (known as P-10 gas) is circulated through the counter at a slow rate during counting and a much faster rate when the counter is opened for exchanging sources, so avoiding contamination of the gas by oxygen or other electronegative gases entering from the atmosphere.

The amplifier used for the proportional counting of β particles must be of the non-overloading type since it has to cope with pulse heights due to energies ranging from near zero to the megaelectronvolt region, the range covered by β particle spectra (Figures 3.6(a) to (d)).

Alpha and beta particle counting

When counting intensely ionising α particles the plateau is reached at a significantly lower polarising voltage than the plateau for counting β particles (Figure 4.10). This is useful (a) because α particles can be counted unaffected by the presence of β particles (though not vice versa) and (b) because the background rate at the α particle plateau is significantly lower than at the β particle plateau (Section 4.6.5).

In 2π proportional counters, the counting gas is sealed in by a thin metallised Mylar membrane, the metallisation being necessary since the membrane forms part of the cathode. The Mylar provides a relatively large entrance window to the counter (see Figure 6.1(a)) which is normally used to detect surface contamination as described in Section 2.6.5. The response of these counters depends on the gas filling (Table 5.1) and on other parameters set by the manufacturer when the counters can be used with few adjustments.

At the moderate pulse amplification used in these detectors compared with that in GM detectors, multiple ionisations by collisions do not spread along the entire anode wire as in GM counters but are confined to small regions and so are more easily controlled. Following a pulse, proportional counters recover within 2 to 4 μs, with a correspondingly short dead time (Section 4.5.2).

5.4 Other detectors and detection methods

5.4.1 A matter of emphasis

This section will serve to discuss liquid scintillation (LS) detectors, microcalorimetry and neutron detectors. Semiconductor detectors will be discussed in Section 5.5. Each of these detectors was designed to serve the character-istics of the detection process of interest. Nevertheless, there is much common ground, notably the need to realise an adequate detection of low-intensity radiation.

Detectors and their associated signal processing equipment are complex systems, the functioning of which need not be known in detail. However, one should be familiar with their working principles which provide helpful guides for the identification of malfunctions.

5.4.2 Liquid scintillation (LS) counting

Introduction

Liquid scintillation counting avoids source self-absorption which is difficult or impossible to avoid when working with solid sources (Section 4.2.4). However, LS counting is subject to other source preparation problems. For example, to effect LS counting of radionuclides in aqueous solutions, the organic scintillant may have to be combined with the aqueous sample and remain a stable and homogeneous mixture until all countrate measurements are completed. Practitioners wishing to employ this technique will find helpful advice in Section 4.3 of NCRP Report 58 (1985), or from radio-chemical suppliers (Section 4.3.3).

Quenching agents

Many problems affecting LS counting are due to the relatively inefficient conversion of the energies of ionising radiations to light quanta taking place in the scintillator molecules. It was noted in Section 4.2.4 that prompt de-excitations of scintillator molecules excited by ionising radiations occur not only with the emission of light quanta but also as molecular vibrations or as thermal energy. In either case counts are lost.

A significant fraction of the light produced by the scintillators is likely to be emitted and/or scattered in directions where it fails to reach the photo-cathode of the PM tube notwithstanding careful design of the detection equipment. Photons are also absorbed in the radioactive solution and in other materials which they have to traverse before releasing photo-electrons at the cathode of the PM tube. These losses of light, known as quenching, are the cause of most inefficiencies which have to be dealt with to obtain reliable LS counting results.

Quenching is also caused by instabilities in the scintillator solutions or emulsions, which are known as cocktails. They are prepared commercially

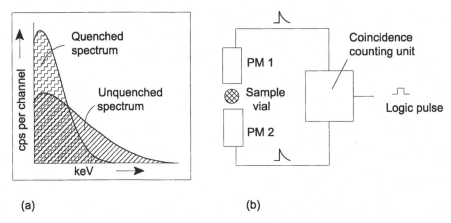

Figure 5.4.   (a) The effect of quenching of an LS spectrum of tritium. (b) An LS spectrometer circuit with two PM tubes connected through a coincidence unit. Coincidences trigger a logic pulse as shown.

according to different formulae depending on requirements. Cocktails have to be handled with care to avoid interference with their chemistry, especially when counting low-energy β particles from ^3H or ^{14}C (Figure 5.4(a)). Low-energy particles are most likely to become subject to quenching or otherwise fail to be counted.

Luminescent effects can be reduced by preparing and keeping vials in the dark prior to counting and by placing the radioactive solution between two PM tubes connected in coincidence (Figure 5.4(b)). Unwanted thermal effects can be reduced by operating the equipment a few degrees above $0\,°C$ (freezing must be avoided). It is also necessary to prevent the access of air to scintillator solutions since oxygen is a significant quenching agent. If suspected to be present, oxygen is removed by slowly bubbling pure nitrogen or argon through the solution prior to counting.

Additional sources of error could be due to noise pulses, chemiluminescence and other spurious light emissions. These effects can often be identified by careful inspection of counting results using a CRO to monitor pulse heights. Some of these effects can be avoided by suitably adjusting the lower level discriminator (LLD). However, 'playing safe' by using too high an LLD setting could lead to significant losses of source counts.

Once sources of error and uncertainties have been dealt with, LS counting can be carried out routinely at the rate of hundreds of vials a day in commercially available automated, computer controlled detection equipment.

Comments on LS counting procedures

LS solutions are significantly less efficient than NaI(Tl) scintillators in the conversion of the energy of ionising radiations into countable pulses. NaI(Tl) crystals require approximately 0.3 keV to produce a countable pulse. For liquid scintillators this energy could exceed 1 keV. The low conversion efficiency accounts for the poorer energy resolution of LS systems when compared with NaI(Tl) detectors. When working with higher energy β particle emitters, e.g. ^{32}P ($E_{\beta max} = 1.7$ MeV) counting efficiencies can reach 96 to 98%. Alpha particles, being strongly ionising, may have to be counted at a lower applied voltage than ^{32}P to avoid multiple pulsing.

5.4.3 Microcalorimetry for routine activity measurements

Counting decays with thermal power

Microcalorimetry can be used to measure energies of the order of microjoules provided they can be converted without loss into thermal energy. Microcalorimetry was employed for radioactivity measurements as early as 1903 and has remained in use to the present time (Mann *et al.*, 1991, Section 4.8). Such measurements are straightforward in principle. A radionuclide R is placed into the microcalorimeter in conditions which ensure that the decay energy is absorbed and measured as thermal power. If the average energy per decay at time T is E_{RT} joules and the thermal power absorbed by the calorimeter and measured at that time is P W (J/s) the activity of R is calculated as $(P/E_{RT})\text{s}^{-1}$.

The method is attractive for measuring large activities of pure β particle emitters with low maximum energies, e.g. ^{3}H or ^{14}C (Section 4.2.4). The particles and the bremsstrahlung are then easily absorbed (even in aluminium (Figure 3.2(b))) and there is no problem with self-absorption effects. However, the very low energies released per decay, 10^{-15} J for ^{3}H, require high activities for sufficiently reproducible results. By contrast, α particles from naturally radioactive elements are emitted with megaelectronvolt energies and, being easily absorbed, they are ideally suited for accurate thermal measurements. Unfortunately, such measurements are rarely practical due to difficulties with unwanted contributions from daughter nuclides (Figure 1.5).

Microcalorimetry has also been employed successfully for measurements of X and γ ray emission rates, and this with γ ray energies up to about 600 keV. Photon energies of this order can be absorbed in many calorimeters with 99% probability by fitting heavy metal to sufficiently attenuate the γ rays and

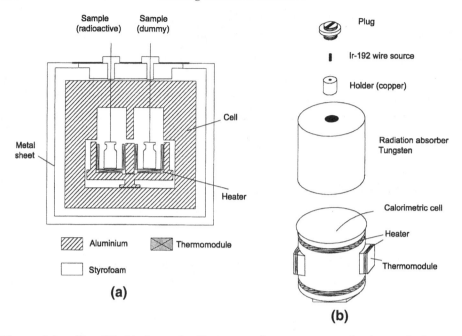

Figure 5.5. Simplified schematic diagrams of components of micro-calorimeters. (a) The twin-cup assembly. (b) Equipment for the activity measurement of an iridium-192 wire (from Genka *et al.*, 1987, 1994).

the bremsstrahlung produced by the absorption of higher energy β particles. In recent years measurements have been made using a twin-cup microcalorimeter where the power liberated by a radionuclide in one of a pair of identically made and carefully insulated cups is balanced by electrical power liberated in its twin.

Microcalorimetry for nuclear radiation applications

The core of a twin-cup microcalorimeter is illustrated in Figure 5.5(a). It provides for relatively large volume cups (each 50 mm diameter × 50 mm high) which are easily manipulated and provide room for sufficient thickness of lead or other metals to absorb relatively high-energy photons, e.g. the bremsstrahlung due to high-energy β particle emitters such as ^{32}P (Figure 3.4(a)). Easy access to the cups ensures that highly active material can be manipulated quickly, so keeping doses to operators low enough not to cause concern (Section 2.6.2).

The calorimeter is set in a temperature controlled space in a console (not shown in the figure) which also contains the necessary instrumentation. Highly sensitive semiconductor thermometry and a carefully stabilised electronically controlled power supply serve to ensure that both cups remain at

Table 5.2. *Radioactivity measurements in a microcalorimeter.*[a]

Nuclide	Sample form	Radiation absorber	Activity (GBq)	Uncertainty (%)
^{3}H	Li–Al alloy in Al		378	± 1
^{14}C	BaCO$_3$ powder	vial	17.9	± 2
^{32}P	H$_3$PO$_4$ in HCl solution	vial + Pb pot	1.80	± 2
^{33}P	solid sulphur	quartz	0.700	± 5
^{35}S	H$_2$SO$_4$ in HCl solution	vial	4.10	± 6
^{90}Y	Y$_2$O$_3$ in polymer	quartz + Cu pot	0.355	± 5
^{153}Gd[b]	pellet in Ti capsule	Cu block	1.17	± 2
^{169}Yb[b]	pellet in Ti capsule	Pb block	3	± 5
^{192}Ir[b]	metal in Pt sheath	W block	0.060	± 4
^{204}Tl[b]	Tl(NO$_3$)$_3$ in NaOH solution	vial	1.32	± 3

[a] The results in this table were obtained as part of a research project carried out at the Japan Atomic Energy Research Institute, Tokai Research Establishment, Japan (see Genka *et al.*, 1987, Genka *et al.* 1994).
[b] These radionuclides emit photons as well as electrons. Approximate figures for the number of γ rays per decay and the highest photon energy are as follows:

gadolinium-153: $\sim 0.5\gamma$/decay, E_{max} 103 keV
ytterbium-169: $\sim 1.4\gamma$/decay, E_{max} 303 keV
iridium-192: $\sim 2.2\gamma$/decay, E_{max} 615 keV
thallium-204: 1.2% Hg KX rays, E_{ave} 80 keV

the same temperature throughout each measurement. To balance the small thermal power requires an equally small electric power and the facility to regulate it to within 0.2 μW or closer.

Results obtained with this instrument over the period 1985 to 1993 were published by Genka *et al.* (1987, 1994) (see Table 5.2). Attention is drawn to the successful activity measurements of both low and high endpoint energy pure β particle emitters (^{3}H and ^{32}P) and of iridium-192 wire (Figure 5.5(b)) which emits photons in the energy range 65 to 612 keV (plus 0.3% at 884 keV) when an activity of only 60 MBq could be measured with an uncertainty of ± 4%. Iridium-192 wires are currently used for medical therapy, being implanted into tumours which the radiations serve to destroy. These procedures require accurate measurements of the iridum-192 activity for which calorimetry measurements are well suited.

Problems which normally arise when working with high-activity materials are eased because all samples can be securely encapsulated. Source self-absorption is helpful rather than a problem, and expensive radionuclides can be completely recovered for re-use.

5.4.4 Neutron detection for scientific and industrial applications

An overview

Methods for the generation of neutrons were introduced in Section 1.3.6. Neutrons are used for numerous applications in science and industry (Section 7.4) and for environmental investigations. They are detected principally by two methods. The first method is the use of proportional counters (Section 5.3.3) with boron trifluoride as the detection medium to measure countrates due to thermal or epithermal neutrons. The neutrons are emitted at mega-electronvolt energies from portable sources (Figure 1.6(a)) and reduced to detectable thermal energy during applications. The second method uses neutron activation analysis.

Proportional counting

Proportional counters can be adapted for neutron detection and their use brings the following two important advantages.

(i)   A portable neutron source (emitting fast neutrons at moderate intensities, see Section 1.3.6) and the proportional counter which is insensitive to fast neutrons, can be mounted together to form a compact, easily transported combination. The detector 'ignores' the fast neutrons from the source but responds efficiently e.g. to the backscattered neutrons which had been thermalised by the surroundings.

(ii) The detectors and their signal processing equipment are reliable and efficient as well as compact and their operation is readily automated for data processing and storage.

The performance of the proportional counter as a detector of thermalised neutrons depends critically on the gas filling. Nuclear reactions in BF_3 occur as $^{10}B(n,\alpha)^7Li$. This makes the absorption of thermal neutrons relatively inefficient in a gas of naturally occurring boron (20% ^{10}B, 80% ^{11}B). Detection efficiencies are greatly improved by employing pure or nearly pure $^{10}BF_3$. But even $^{10}BF_3$ will only respond to neutrons of energies not much greater than 0.025 eV. If it is necessary to also detect higher energy neutrons, the $^{10}BF_3$ has to be replaced by helium-3 which responds to neutrons with energies up to about 1 eV.

Neutron energies are classified depending on the ways the neutrons are used. A frequently employed classification is given by L'Annunziata (1998, Section 1.2.1). In this book neutron energies are classified as follows:

thermal neutrons (near room temperature) approx. 0.025 eV
epithermal neutrons up to 1 eV
fast neutrons exceeding 10 keV

If proportional counters containing $^{10}BF_3$ or ^3He have to detect fast neutrons, the counters are embedded in 10 to 12 cm thick paraffin, a hydrogenous material which moderates the fast neutrons down to detectable thermal energies. The paraffin is covered with a thin sheet of cadmium which absorb all thermal neutrons reaching the surface of the paraffin, the lower cut-off energy being about 0.05 eV.

Measurements using high-intensity neutrons

Neutrons are generated in research reactors at flux densities up to 10^{15} n/ $(cm^2\ s)$ and of energies up to about 8 MeV with an average energy near 2 MeV. Their high energies are at once moderated, resulting in a neutron spectrum where the large majority of the neutrons have thermal energies, with smaller proportions extending to intermediate and high energies.

The high-intensity thermal neutron flux in research reactors is used, among other things, for the irradiation of samples of materials that are known or expected to contain rare minerals or unwanted impurities. Following activations, the samples are placed in a γ ray spectrometer when the activated nuclides can be identified by their spectra, a procedure known as neutron activation analysis. A few details will be provided in Sections 6.4.5 and 7.4.4.

5.5 An introduction to semiconductor detectors

5.5.1 A few historical highlights on energy spectrometry

Small, semiconducting silicon detectors have been used for energy spectrometry of α particles and electrons for several decades. These detectors are a few millimetres thick (Figure 5.6) and can be made pure enough to ensure a sufficiently high resistivity and so a satisfactory performance even at room temperature (Section 5.5.2). Since the resistivity of semiconductors increases as their temperature decreases (in contrast to electrical conductors), their resolution improves with decreasing temperature of the detector as shown in Figure 5.7.

The small size of silicon detectors and their relatively low density (2.33 g/cm^3) makes them unsuitable for γ ray spectrometry. So far, it is only germanium crystals ($\rho = 5.4$ g/cm^3) at liquid nitrogen temperatures that can be used as detectors for high-resolution γ ray spectrometry, up into mega-electronvolt energies (Heath, 1974, Debertin and Helmer, 1988, Section 2.1.3). Collections of γ ray energies and intensities emitted from frequently used radionuclides will be found in NCRP (1985, Appendix 3). For more

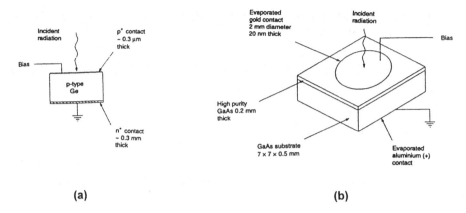

(a)   **(b)**

Figure 5.6. (a) A cross section through a planar high-purity p-type germanium detector (HPGe) (Debertin and Helmer, 1988, Figure 2.1). (b) A gallium arsenide semiconducting crystal detector, dimensions enlarged for clarity (Alexiev and Butcher, 1992).

detailed data see Lederer and Shirley (1978) or Firestone and Shirley (1996), though the latter could be too detailed except for specialists.

Going back to scintillation detectors, solid scintillation spectrometers based on NaI(Tl) crystals came into operation during the late 1940s when they opened up the use of γ ray emitting radionuclides for a large and still growing number of applications (Heath, 1964).

However, the energy resolution attainable with NaI(Tl) detectors is inadequate for many tasks. This is illustrated by comparing the ^{241}Am photon spectrum in Figure 3.1 taken with a NaI(Tl) detector with the spectrum of ^{241}Am in Figure 5.7(b) taken with a semiconducting compound detector, also at room temperature (Alexiev *et al.*, 1994). The improvement in energy resolution is greater still if comparisons are made with the spectrum taken at a lower temperature (Figure 5.7(a)).

5.5.2 *Characteristics of germanium and silicon detectors*

High-resolution semiconducting detectors are crystals of silicon or germanium. The latter have to be used at liquid nitrogen temperatures (about $-196\,^{\circ}$C), but small silicon detectors have long been employed for particle spectrometry at temperatures up to about $60\,^{\circ}$C. Two characteristics which permit sufficiently pure semiconductor crystals to function as high-resolution detectors of ionising radiations, even at and above room temperature, are a relatively wide band gap energy and a high resistivity at low temperatures which can only be achieved by highly pure crystals (see below).

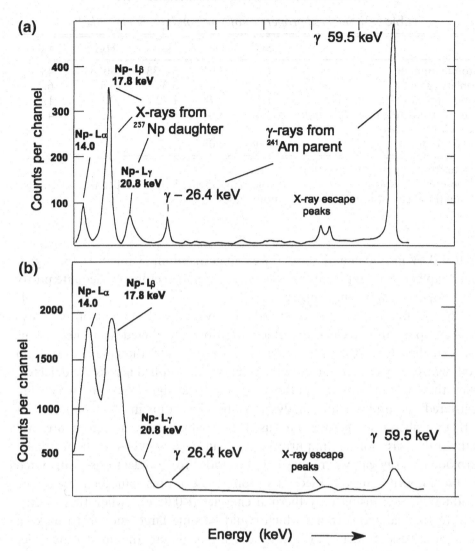

Figure 5.7. Photon spectra following the decay of ^{241}Am, detected with a GaAs semiconducting crystal (a) at 122 K, bias 100 V and (b) at 295 K (room temperature), bias 84 V. The photon spectra are due to the daughter, ^{237}Np. Note the differences in energy resolution as well as in the counts per channel (Alexiev and Butcher, 1992) (also see Figure 3.1).

Whereas electronic states in free atoms are sharply defined (Figure 3.16(a)), those in crystals of semiconductors are broadened into bands that are characteristic of the crystal as a whole. Electrons are located in the valence band, which is separated from the conduction band by a region known as the band gap in which no electronic states are allowed. This region has a width of

Table 5.3. *Selected properties of semiconducting crystals.*[a]

	Si	Ge	GaAs	HgI$_2$	CdTe
Atomic number	14	32	31, 33	80, 33	48, 52
Density (g/cm^3)	2.3	5.4	5.3	6.4	6.2
Band gap (eV)	1.12	0.74	1.42	2.1	1.5
Energy to create an electron–hole pair (eV)[b]	3.6	3.0	4.5	4.2	4.4
Fano factor	0.08	0.08	–	–	–

[a] Data from Sumner *et al.* (1994).
[b] These energies apply for electron–hole pairs created by α or β particles emitted during the decay of normally used radionuclides.

up to 1.3 eV in silicon and about 0.7 eV in germanium (Table 5.3), the higher band gap being an important reason why silicon detectors of adequate purity can be operated at room temperature.

When germanium crystals are irradiated by γ rays, the energy of the γ rays is taken up by atomic electrons which are promptly ejected from their atoms and can then cross the approximately 1 eV gap into the conduction band from which they are collected by the electric field applied across the detector. They then account for the pulses of charge (the signal current) which are registered by a multichannel analyser or other recording instruments.

If the silicon or germanium used to manufacture detectors are not sufficiently pure, ionised impurity atoms can occupy states in the normally forbidden energy gap where they act as a bridge for the electrons contributed by the impurity atoms. Electrons from these atoms can then reach the conduction band using only thermal energies (<0.05 eV) when they contribute to the leakage currents which could become large enough to make a detector useless for its task. Effects due to impurities are the smaller the wider the band gap. Compound semiconductor crystals (Section 5.5.5) can be readily operated at room temperature and higher, because they have significantly wider band gaps than even silicon (Table 5.3).

Signal currents commonly have to compete against leakage currents due to bulk and surface effects. With high-purity germanium detectors at liquid nitrogen temperature, leakage currents can be kept sufficiently small notwithstanding the high field strength of about 1000 V/cm which is needed to collect the signal current in germanium crystals.

5.5.3 *Lithium drifted and high-purity germanium detectors*

The first procedure which produced germanium crystals of acceptable purity and size for γ ray detection was to drift ionised lithium atoms into the crystal. They neutralise the impurity atoms (mainly boron), so preventing them from accepting or donating electrons. The method is complex but keeps leakage currents sufficiently low to permit the manufacture of relatively large-volume high-resolution detectors known as lithium drifted germanium detectors, Ge(Li). These detectors, however, suffer from two serious shortcomings. First, the drifting process leaves a dead layer at the surface blocking all photons with energies below 30 or even 40 keV and significantly attenuating higher energy photons up to about 100 keV. Second, the crystals have to be stored as well as used at liquid nitrogen temperature to prevent the lithium atoms from drifting into positions where they lose their effectiveness.

Eventually germanium crystals could be produced with only about 10^{10} effective impurity atoms per cubic centimetre or even less (Debertin and Helmer, 1988, Section 2.1). It then became possible to produce intrinsic, p-type, high-purity germanium (HPGe) crystals (Figure 5.6(a)). They have to be used at liquid nitrogen temperature but can be stored at room temperature without losing their effectiveness. High-purity germanium detectors are replacing lithium drifted detectors. The latter are in many respects satisfactory, but to keep them continuously at $-196\,°C$ is costly and troublesome.

The effect of the dead layer on the surface of Ge(Li) detectors can be seen when comparing the spectra shown in Figures 3.9 and 3.14. In Figure 3.9(b), a spectrum of ^{133}Ba, the dead layer absorbed the X rays which are prominent in the ^{133}Ba spectrum due to a NaI(Tl) detector shown in Figure 3.9(c). In Figure 3.14(a) (a europium-152 spectrum), little is left of the around 40 keV KX rays which are emitted with more than twice the intensity of the 122 keV γ rays.

Figures 3.14(b) and (c) show the improvements when the cobalt-57 spectrum is obtained with a HPGe detector. The latter displays the 14 keV γ rays, the about 6.5 keV KX rays emitted by the daughter (iron), and the peaks due to summing of iron KX rays and γ rays. The majority of HPGe detectors are either planar shaped (Figure 5.6(a)), or co-axial, a cylinder drifted radially from a 10 to 15 mm diameter concentric bore. Detector windows can be made thin enough to clearly detect 3 keV photons.

5.5.4 *Further comments on silicon detectors*

Silicon detectors are produced as either ion implanted or surface barrier or lithium drifted detectors. Ion implanted silicon detectors have low leakage currents even at room temperature and can be operated at temperatures up to 60 °C. However, energy resolution improves with decreasing temperature which can be effected using Peltier cooling.

A 0.1 mm thick silicon detector will take care of most requirements for α particle spectrometry with numerous applications for environmental investigations. Crystals for β particle or electron spectrometry are made up to 5 mm thick and about 20 mm diameter. The low density and low atomic number of silicon ($\rho = 2.33$ g/cm^3 and $Z = 14$) ensures a low backscatter intensity, a useful characteristic for β particle spectrometry measurements.

Ion implanted silicon detectors operated at -30 °C can be used for high-resolution X ray spectrometry but only at relatively low energies. They are almost 100% efficient within the energy range 2 to about 20 keV. Their efficiency is still usefully high up to about 25 keV but then drops rapidly to very low values.

5.5.5 Detectors made from crystals of semiconducting compounds

The characteristics of commercially available semiconducting compounds used as photon detectors were published by Sumner *et al.* (1994). Some of these data are reproduced in Table 5.3, which lists properties of three compounds, gallium arsenide (GaAs), mercuric iodide (HgI$_2$) and cadmium telleride (CdTe), along with the properties of silicon and germanium for comparison.

The useful features of crystals of semiconducting compounds have to be set against their small size. To retain their useful properties as radiation detectors they cannot at present be made larger than small fractions of a cubic centimetre (Alexiev and Butcher, 1992).

However, small-sized high-resolution detectors, which can be operated above room temperature, can serve for medical applications and for other applications where γ ray spectrometry and dosimetry are used in confined spaces. The polarising voltage across these small detectors rarely has to exceed about 100 V, which is much more compatible with many applications than the much higher polarising voltage needed for germanium detectors.

Figure 5.7 shows photon spectra of americium-241 obtained at -150 °C and at room temperature with a GaAs crystal detector similar to that shown in Figure 5.6(b) (Alexiev and Butcher, 1992). The energy resolution of the KX and γ ray peaks (they are neptunium-237 peaks) is, not surprisingly, better at -150 °C than at room temperature. However, notwithstanding its small size, the semiconducting compound detector yields a more informative spectrum at room temperature than the much larger NaI(Tl) detector shown in Figure 3.1.

5.5.6 Energy resolution

The energy resolution of a pulse height peak is defined as the full width of the peak in kiloelectronvolts at half its maximum height (FWHM). It is their excellent energy resolution which makes semiconducting detectors the extremely useful instruments they are. For gaseous detectors or semiconducting crystals, energy resolution depends primarily on the amounts of energy needed to create ion pairs in the material of the detector and on the statistics of these processes.

Ionisations of the gas in a proportional counter produce negatively charged electrons and positively charged gas molecules. Ionisations of atoms in semiconducting crystals produce electron–hole pairs. Holes can move through the crystal like electrons. They can be pictured as a positively charged elementary particle resulting from the removal of an electron.

The average energy needed to create electron–hole pairs in silicon or germanium is only about 3 eV. In contrast, to create ion pairs in a gaseous detector requires on average approximately 30 eV (Table 5.1). Hence, the absorption of a given amount of photon energy in a semiconducting detector results in a ten times larger number of charge carriers than does the absorption of the same energy in a gaseous detector, so leading to a greatly improved energy resolution (see below). Compound semiconducting crystals require up to 4.5 eV to create an electron–hole pair (Table 5.3). This is still a substantial improvement on gaseous detectors. The resolution of scintillant detectors, e.g. NaI(Tl) crystals, depends on different processes (Sections 4.4.1 and 6.4.4).

Electron–hole pairs are generated by several mutually independent processes, all subject to random variations as are decay rates (Section 2.2.1). This is allowed for by the Fano factor F, Table 5.3, but only for silicon and germanium. Values of F are similar for the semiconducting compounds but, for reliable results, should be measured in each case. Assuming this has been done, the FWHM value of a full energy peak (w keV), obtained with a semiconducting detector is given by:

$$w = k(F \times E_\gamma \times p)^{1/2}, \qquad (5.3)$$

where k is a constant which equals 2.36 for Gaussian peaks (Figure 6.11), F is the Fano factor for the detector material, E_γ keV is the energy of the detected γ rays and p keV is the average energy needed to create an ion pair in the detector material.

Equation (5.3) has to be used with care since it is based on the assumption that the FWHM of a peak is unaffected by other factors. This is rarely so in practice. Almost invariably there are contributions, for example, from leakage currents, electronic noise and charge trapping leading to incomplete collection of electrons and/or holes (Section 5.5.5). One or all of these effects are difficult to avoid with all of them contributing to peak broadening. However, careful work can commonly ensure that peak broadening is minimised as far as possible. Other formulae for the FWHM will be introduced in Section 6.4.4.

5.5.7 *A postscript on semiconducting detectors*

The satisfactory operation of semiconducting detectors depends on the complete collection of electrons and holes produced by the ionising radiation while avoiding or neutralising pulses of charge due to other processes. This applies in particular if one wants to obtain optimum results from high-resolution detectors. They should respond only to wanted signals, i.e. those caused by photons emitted from the radionuclide of interest. If that is so, the spread of pulse heights is likely to be minimised since it is pulses from "foreign" sources that are principally responsible for widening a pulse height distribution and so causing an unwanted increase in the FWHM value of the peak which semiconductor detectors are used to minimise.

Readers seeking to obtain information on high-performance detectors and associated signal processing equipment should get in touch with manufacturers, all of whom are continually engaged in intensive research to improve the quality of their products while lowering costs.

Chapter 6

Radioactivity and countrate measurements and the presentation of results

6.1 An introduction to radioactivity measurements

6.1.1 Problems

Many comments in preceding chapters made it clear that accurate radio-activity measurements require specialised instruments and attention to numerous details, in particular an adequate knowledge of the decay data of the radionuclides of interest.

Highly accurate radioactivity measurements are rarely of interest outside standard laboratories, but two facts require attention: (a) all work with radioactivity relies ultimately on internationally established activity standards (Section 2.2.1), and (b) laboratories working with radioactivity must have facilities for at least moderately accurate radioactivity measurements since radiation protection authorities will not permit radioactive materials to be used unless their activity is known with sufficient accuracy, often within ±10%.

Since most nuclear decays are signalled by the emission of either an α or a β particle, the measurement of these decay rates could be expected to be straightforward: one counts the emitted particles for a known time and states the result as decays per second or becquerel (see Eq. (2.1)).

To proceed in this way could cause serious errors. For example, there are the β particle decays via excited states de-excited by conversion electrons (Section 3.6.2) which the detector treats like β particles, adding them to their number. Furthermore, to account for all emitted particles, counting must be done in exactly 4π or another accurately known geometry requiring appropriate detector arrangements and the source must be thin enough to permit all particles emitted from within its atoms to escape to be counted. In addition, there are uncertainties due to imperfections of the signal

147

processing equipment (Section 4.5), the possible presence of unwanted radiations (Section 4.6) and ambiguities in interpretation of results (yet to be discused).

Furthermore, detectors record countrates due to the radiations that triggered them, not decay rates in the radioactive source. To obtain decay rates, detectors have to be calibrated using standardised sources with the appropriate standard for each radionuclide (Section 2.2.1) when all measurements have to be made in the source–detector geometry used for the calibration and have to be corrected as needed.

6.1.2 A role for secondary standard instruments

Radioactivity standards are normally obtained commercially (Section 4.3.3). However, they are expensive and often of limited lifetime. When radioactivity measurements are required routinely, there are two alternative methods that may be preferable: radioactivity measurements with secondary standard instruments (SSIs, Section 2.2.3), or, if accuracies need only be moderate, the purchase of what are here called working standards.

In most countries it is now mandatory for the activity concentration of samples of radionuclides supplied commercially to agree with the stated values within ±10%. This means that purchased solutions, received in sealed containers, could be used as working standards since a ±10% uncertainty in the results of countrates is adequate for many routine applications. However, when a radioactive solution is used as a working standard, it is essential to ensure that its radioactivity concentration remains unaffected during storage or when aliquots are withdrawn (this must be done quantitatively).

SSIs serve to verify the activities of X and γ ray emitters and also to monitor the reproducibility of activity measurements which can be done for any radionuclide emitting photons of energies within the range of the SSI. The most widely used SSIs are the $4\pi\gamma$ pressurised ionisation chambers and the γ ray spectrometer. Before discussing these instruments, it will be helpful to introduce a few comments on the establishment of radioactivity standards since, as noted earlier, all measurements of radioactivity ultimately rely on these standards.

The concluding sections of this chapter will discuss methods for estimating random and systematic uncertainties in the results of measurements in nuclear radiation applications followed by a discussion of the role of statistical theories for uncertainty estimates.

6.2   Comments on the preparation of radioactivity standards

6.2.1   Problems with beta particle emitters

To effect accurate radioactivity measurements requires sufficient competence to deal with problems such as those discussed in Section 6.1.1. When using β particle emitters, which is the case for most applications, it must be possible to account for all decays in a source even when sources are insufficiently thin so that a percentage of the β particles is unable to escape.

This problem was solved (or almost so) by the late 1950s, but initially only for radionuclides that decay with the emission of γ rays in coincidence with the β particles, as applies, for example, to ^{60}Co (Figure 3.4(b)). The method became known as 4πβ–γ coincidence counting (Campion, 1959). Furthermore, in order to obtain sufficiently accurate results (within ±0.2%), sources of radioisotopes that emit β and γ rays in coincidence should be made thin enough to permit at least 90% of the β particles to escape to be counted. The reasons for this will be stated presently.

6.2.2   Accurate radioactivity measurements

By the mid-1950s many radionuclides could be prepared as very thin sources which could be counted in newly developed windowless 4πβ proportional gas flow counters (Figure 6.1(a)) with individual results reproducible to within ±0.10%. However, results of activity measurements made on thin 4π sources prepared from samples of solutions of several radionuclides distributed to standard laboratories in different countries always differed by more than the attainable reproducibilities. This could be shown to be due mainly to errors in source self-absorption corrections which had to be estimated by subjective procedures. It was thanks to the invention of 4πβ–γ coincidence counting that the activity of a source emitting β and γ rays in coincidence need no longer be obtained as a subjective estimate but could be calculated from measured countrates. The method will be demonstrated using ^{60}Co and the circuit shown in Figure 6.1(b). This is a greatly simplified version of what is normally employed but adequate for the present purpose.

The 4π proportional counter (labelled 1 in Figure 6.1(b)), is in contact with a γ ray counter (2), usually a NaI(Tl) crystal. The electronic instrumentation includes the amplifiers and pulse processors (3, 5) and a coincidence unit (7) which is triggered by the coincident arrival (within an adjustable number of microseconds) of signals from the two detectors.

Sources of ^{60}Co can be made thin enough to have a high β particle efficiency

(a)

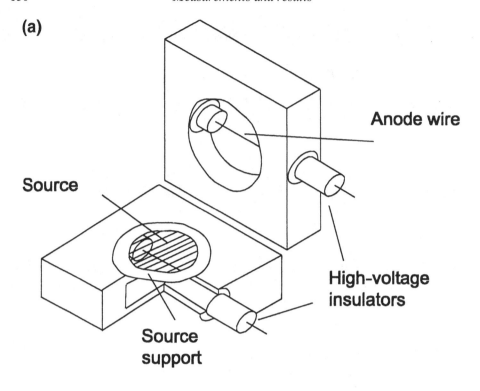

(b)

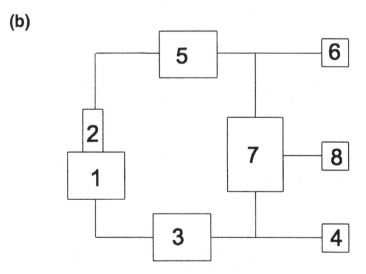

Figure 6.1. (a) A 4πβ thin wire windowless proportional counter. The cavities are each 50 mm in diameter and 20 mm deep. (b) The circuit diagram for 4πβ–γ-coincidence counting (see Section 6.2.2 for details).

($\varepsilon_\beta > 90\%$), a preferred condition to obtain accurate results (see below). One counts N_β cps in the β particle counter, N_γ cps in the γ ray counter and N_c cps when radiations from the two counters arrive in coincidence. The counters are numbered 4 (β), 6 (coincidence) and 8 (γ) in Figure 6.1(b).

Applying all the necessary corrections (which cannot be detailed here for the sake of brevity), the disintegration rate N_0 Bq of the ^{60}Co in the source and at the time of the measurement can be expressed in terms of observed countrates when:

$$N_0 = \varepsilon_\beta N_\beta \times \varepsilon_\gamma N_\gamma / \varepsilon_\beta \varepsilon_\gamma N_c \qquad (6.1)$$

The counting efficiencies,= ε_β, ε_γ and $\varepsilon_\beta \times \varepsilon_\gamma$ cancel to a close approximation, so that N_0 values can be obtained from Eq. (6.1) as $N_\beta \times N_\gamma / N_c$. However, if ε_β is not high enough the γ rays from the source (they escape readily in contrast to the coincident β particles) could trigger the β counter so causing an error. Fortunately, this error can be kept small (<0.2%) provided $\varepsilon_\beta > 0.90$.

The coincidence method was soon extended to radionuclides with complex decay schemes, eventually using computerised control of the instrumentation. The principles of the method are described in NCRP (1985), Section 3.2.2, or can be electronically downloaded (see Table A3.1 in Appendix 3).

6.3 4πγ pressurised ionisation chambers

6.3.1 Introduction

Since γ ray emitters are of primary interest for industrial applications, the fact that ionisation currents due to γ rays can be measured relatively simply and accurately in pressurised 4πγ ionisation chambers is of considerable interest. For accuracies within ±3%, calibrations have to be effected using standards of radioactivity when the chambers become secondary standard instruments (SSIs). For work within ±3 to ±10%, one could use purchased solutions as standards as described in Section 6.1.2. Given the linearity of the response of SSIs as functions of the γ ray energy (Figure 2.1), calibration constants of many γ ray emitters can be estimated within about ±10% from data such as those presented in Table 6.1.

6.3.2 Two types of 4πγ pressurised ionisation chambers

4πγ pressurised ionisation chambers are supplied commercially either as general purpose instruments, known simply as gamma ion chambers (Figure 6.2), or

Table 6.1. *Data for the operation of 4πγ pressurised ionisation chambers:*

(a) Parameters recommended for the capacity charging method.[a]

Working medium	argon or nitrogen, ~2 MPa
Plug-in capacitances (pF)	1000, 3000, 10 000, 30 000, nominal
^{226}Ra activities (in equilibrium with daughters)	MBq 0.35, 0.70, 3.5, 17.5, nominal
Stray capacitance (pF)	<100
Background and leakage currents (pA)	<0.25

(b) Results obtained with a 2 MPa argon chamber.[b]

Radionuclide	$T_{1/2}$	Calibration factor(pA/MBq)	$(E_\gamma)_{ave}$ (MeV)	$\sum(E_\gamma f_\gamma)$ (MeV)
^{22}Na	2.602 y	59.2	0.78	2.2
^{51}Cr	27.7 d	1.01	0.32	0.032
^{54}Mn	312.2 d	23.3	0.83	0.84
^{59}Fe	44.51 d	31.0	1.20	1.20
^{60}Co	5.269 y	63.8	1.25	2.5
^{88}Y	106.6 d	65.4	1.38	2.7
^{99m}Tc	6.007 h	5.85	0.14	0.13
^{133}Ba	10.54 y	12.6	0.36	0.48
^{137m}Ba	2.55 m	16.3[c]	0.66	0.56
^{198}Au	2.694 d	12.5	0.41	0.40

[a] These data are for guidance only. Laboratories working with sources which have similar activities could work satisfactorily with a single capacitance and radium source. It may also be necessary to accept a higher stray capacitance and leakage current.

[b] Calibration results could differ by up to ±25% in other nominally 2 MPa argon chambers (compare data in Figures 2.1 and 6.5). The radionuclides are among those frequently used for applications.

[c] This constant was measured using a ^{137}Cs source.

as dose calibrators, also known as radionuclide calibrators (see photograph in Figure 6.3). The following are similarities and differences between these instruments:

(1) Both are strongly made, relatively large volume pressurised ionisation chambers, 35 to 45 cm high, 12 to 18 cm in external diameter. They are operated in the saturation region (Section 5.2.1).

(2) Their large dimensions and mode of operation ensures that the ionisation current due to small sources in their central region, say 10 ml capacity ampoules or smaller, remains constant within 1 to 2% when the sources are moved axially by a few centimetres (Figure 6.4(b)) or radially by a few millimetres (Figure 6.4(c)).

(3) The inner diameter of the re-entrant well of dose calibrators is 5 to 8 cm that of general purpose chambers is often no more than 2.5 cm.

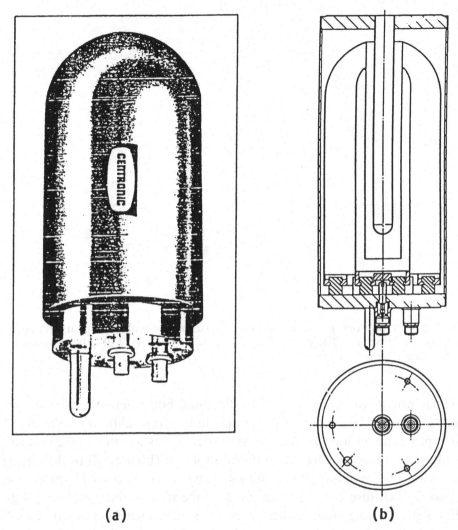

(a)   **(b)**

Figure 6.2. A 4πγ pressurised ionisation chamber. (a) outside its lead cover and (b) in cross section. See Table 6.2(a) for dimensions.

(4) The signal processing equipment used to measure the ionisation current of general purpose chambers is supplied, adjusted and calibrated by the user, whereas the instrumentation of dose calibrators is supplied and calibrated by the manufacturer.

6.3.3 Dose calibrators

Dose calibrators are the instruments of choice for nuclear medicine departments and other laboratories routinely employing a range of different radio-

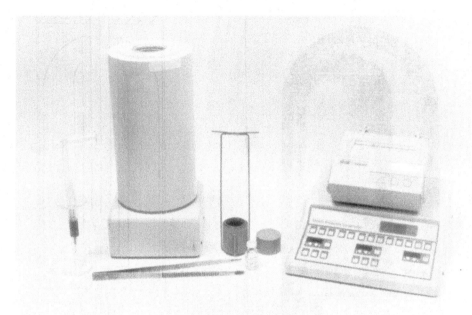

Figure 6.3. A photograph of a radionuclide calibrator with its signal processing equipment and printer. The plastic inset serves to position vials or other radioactive sources where required.

nuclides which may come in sufficiently small but differently shaped containers. To read the activity of the sample in the dose calibrator, users select the appropriate push button or potentiometer setting as listed in the manual, so adjusting a resistor network in the control unit (Figure 6.3) to the setting required for the radionuclide of interest. The voltage drop across the chosen resistor is then displayed as the activity of the radionuclide on a four digit LED (light emitting diode) either in curies or in becquerels as required, with measurements often taking no longer than 20 to 30 s.

The manufacturer's calibration is subject to stated conditions. Changes in these conditions require appropriately changed calibration factors. It is also essential to make all necessary corrections and to guard against errors due to bremsstrahlung or radionuclidic impurities (Section 4.6). Precautions have to be particularly carefully observed at X or γ ray energies below about 150 keV when the ionisation current due to these photons becomes an increasingly sensitive function of the position, density and shape of containers and sources.

To minimise effects due to backscattered and background radiations the ionisation chambers are commonly shielded by 2 to 3 cm thick lead, which also shields operators from doses due to high-activity sources.

Dose calibrators are pressurised ionisation chambers and as such are rugged and stable instruments. Nevertheless, the stability of their response and absence of contamination should be regularly checked using a long-lived source and the manufacturer's calibration should also be verified at least once per year. Recently marketed instruments are offered with computerised instrumentation and facilities to record information about the radioactive samples and their origin. Readers are referred to the catalogues of manufacturers for further information (Section 4.3.3).

6.3.4 General purpose pressurised ionisation chambers

Their role as precision instruments

If accurately calibrated (Sections 6.4.3), pressurised ionisation chambers are routinely employed as SSIs in standard laboratories and in many industrial laboratories.

Ionisation currents that are large enough to produce a clearly measurable response in the chamber can be measured within ±0.10% reproducibility. This is done by effecting the measurements in identically made flame-sealed ampoules or other standard containers using identical volumes of aqueous solutions, all of the same density. The centre of the solution should be situated in the centre of the region of maximum sensitivity (Figures 6.4(b) and (c)). Conditions can be more flexible if it is sufficient to work with more moderate accuracies.

For many routinely made measurements, ionisation currents are calculated from the voltage drop across a high-quality stable resistor, as applies to dose calibrators, the operating voltage being commonly between 500 and 800 V. If accuracy is important it is preferable to measure ionisation currents using the capacitor charging method when the current from the ion chamber is fed into an electrometer–amplifier, the output of which is made to charge a high-quality plug-in capacitor (Table 6.1(a)) selected to ensure that charging times are neither too long nor too short for accurate timing (Santry *et al.*, 1987).

It is also advisable to measure the current due to the source of interest in tandem with that due to a high-quality, long-lived reference source, preferably ^{226}Ra, and express all results as the ratio of these two currents which should be preferably within a factor of two. One then maintains a series of reference sources and selects the one closest in activity to the source of interest (Table 6.1). The ratio method greatly helps with the realisation of reproducible results because it compensates for small drifts in the response of the instrument and helps to detect radionuclidic impurities.

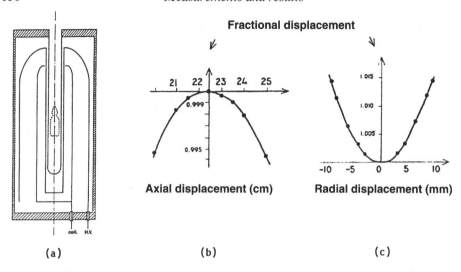

Figure 6.4. (a) A cross section through a pressurised ionisation chamber. (b) Changes in the ionisation currents (fractional displacements) due to small axial and (c) small radial displacements of a point source from the position of maximum sensitivity (Rytz, 1983).

Activity calibrations

Activity calibrations relate the observed specific ionisation current (commonly expressed in picoampere (pA)) per megabecquerel (MBq) activity of the radionuclide in the chamber to the energy of the emitted γ rays (Figure 6.5) (Hino *et al.*, 1996, Kawada *et al.*, 1996). Values of this ratio for frequently used radionuclides measured in a typical pressurised ionisation chamber (2 MPa of argon) are listed in Table 6.1 together with data useful for the efficient operation of these chambers. Table 6.2 lists dimensions of these chambers and characteristics of nitrogen and argon which are the working media normally employed.

If ionisation currents are measured with the capacitor charging method, the measured quantity is the time Δt (seconds) needed for the current I (picoampere) to charge a stable, high-quality capacitor of capacitance C (farad), causing the voltage across C to increase through a predetermined range $V_2 - V_1 = \Delta V$, usually of the order of a few volts. The charging time is then given by $\Delta t = C \times \Delta V/I$ and so is shorter the larger the current. On reaching V_2 the capacitor is discharged and the charging and discharging cycle is repeated automatically in otherwise identical conditions when the average value of Δt is calculated together with its standard deviation (Rytz, 1983).

Let us suppose the standard source used for the calibration is a ^{60}Co solution in a flame-sealed ampoule supplied by the BIPM (Rytz, 1983,

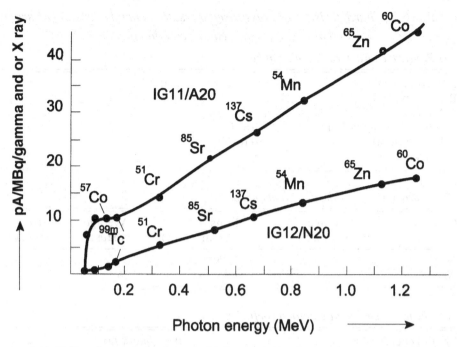

Figure 6.5. Calibration graphs for two ionisation chambers using pressurised argon (A20) and pressurised nitrogen (N20). For ionisational currents due to radionuclides that emit more than one γ ray per decay the photon energy is the weighted linear average (reproduced from Hino *et al.* (1996), see also Kawada *et al.* (1996)).

Figure 6.4(a)) certified as 4.10 MBq at the time of the measurement when the ionisation current, corrected as required (Sections 4.5 and 4.6), was measured as 296.5 pA. The specific ionisation current per megabequerel is then 296.5/4.10 = 72.32 pA/MBq, plotted as a function of the average energy of the two ^{60}Co γ rays, (1/2)(1173 + 1332) = 1253 keV, as shown in Figure 6.5 (see also the works of Hino *et al.* (1996) and Kawada *et al.* (1996)). These figures are of course subject to uncertainties which should always be estimated and shown.

Table 6.1(b) lists calibration factors and $\sum(E_\gamma f_\gamma)$ values (see Equation (2.6), Section 2.7.2). This is the sum of the products of the energies and fractions of all γ rays emitted by the stated radionuclide (as just shown for ^{60}Co) For any two radionuclides R_1 and R_2 the ratio $\sum(E_\gamma f_\gamma)_{R1} / \sum(E_\gamma f_\gamma)_{R2}$ is approximately equal to the ratio $(pA/MBq)_{R1}/(pA/MBq)_{R2}$, which can be used to calculate pA/MBq values for radionuclides of known and sufficiently similar decay schemes.

For example, If $(pA/MBq)_{Na22}$ is calculated from the data for ^{54}Mn using the method outlined above, the result is within 5% of the calibrated value.

Table 6.2. *Characteristics of pressurised ionisation chambers*[a] *and of argon
and nitrogen when used as detection media:*

(a) Physical characteristics of chamber.

Detection media	nitrogen or argon at 2 MPa
Re-entrant well	
Material	mild steel
Internal diameter (mm)	30 to 70
Length (mm)	280 to 320
Thickness (mm)	1.5

(b) Characteristics of the detection media.

Gas	Atomic number	Density, $\rho(g/cm^3 \times 10^3)$[b]	Average energy to create an ion pair (eV)
N_2	7	1.25	35
Ar	18	1.77	26

(c) The mass–energy attenuation coefficient, μ_{en}/ρ[c].

Photon energy	$10^2\mu_{en}/\rho(cm^2/g)$	
keV	N_2	Ar
20	37.5	810
50	3.2	49
100	2.2	7.3
200	2.7	3.0
500	3.0	2.7
1500	2.6	2.3

[a] Readers are referred to suppliers of these instruments (see e.g. Section 4.3.3) for information on currently available designs.

[b] These densities apply at 0 °C and 1.013×10^5 Pa pressure.

[c] This coefficient measures the fraction of photon energies transformed in the gas to the kinetic energies of the ions collected as the ionisation current (Hubbell, 1982). The estimated uncertainty is about ±5%.

For radionuclides with less similar decay schemes the predicted result would be less accurate, but the stated procedure is likely to provide a useful first approximation to the required datum.

Figure 6.5 shows calibration graphs for both 2 MPa argon and 2 MPa nitrogen. Argon yields the higher specific ionisation currents but nitrogen yields a more linear response which is important for accurate interpolations. The relevant characteristics of nitrogen and argon are listed in Table 6.2.

The calibration graph

The calibration graph for a pressurised ionisation chamber is the line of best fit through as many calibration points as appear necessary to arrive at the required accuracy (Figure 6.5). The graph is used to measure not only the activities of the radionuclides used for the calibration, but also for other radionuclides (though often with reduced accuracy). It is of course necessary that the energies of X and γ rays emitted by other nuclides are within the range of the calibration graph and that the source–detector geometry is reproduced with sufficient accuracy.

Calibration graphs for chambers which employ argon as the working medium often show a discontinuity in the gradient near 150 keV (see also Figure 2.1). The short upwards slope is due to the increasingly steep rise of the attenuation coefficient of argon at photon energies below 200 keV as shown in Table 6.2, the rise being very much less steep in nitrogen. However, photon attenuation in the metal used to construct the chamber soon out-weighs effects due to the increasing rate of interactions in the gas, as seen in Figure 6.5.

6.4 Gamma ray spectrometers and gamma ray spectrometry

6.4.1 Towards multi gamma ray spectrometry

Section 5.5 introduced high-precision γ ray spectrometry, referring briefly to NaI(Tl) detectors and in more detail to semiconducting detectors and multi gamma ray spectrometry. This section will take up these subjects and, in particular, full energy peak calibrations of γ ray spectra. This will be followed by a comparison of the performance of γ ray spectrometers with pressurised ionisation chambers when used as SSIs.

Multi gamma ray spectrometry has been greatly advanced by the con-tinuing development of computerised multichannel analysers (MCAs). They operate with built-in electronics that sort the shaped and amplified pulses from the detector as functions of their pulse heights expressed as channel numbers. Currently supplied MCAs provide for between 2^{11} and 2^{13} channels (2048 to 8192 channels) commonly labelled as multiples of 1000. MCAs used with semiconducting detectors should have at least 2000 channels. The resolving power of NaI(Tl) detectors is adequately served by 512 channels.

Gamma ray spectrometry serves to identify radionuclides and their activ-ities from the spectra of emitted γ rays such as the two barium-133 spectra due to a high-resolution and a NaI(Tl) detector (Figures 3.9(b) and (c)) and

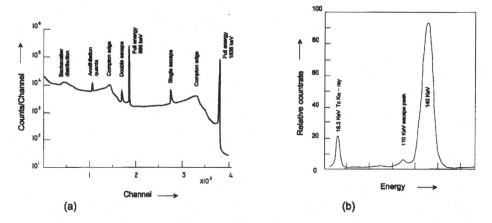

Figure 6.6. (a) The single and double escape peaks from the ^{88}Y 1836 keV full energy peak and also the peak due to the 511 keV annihilation γ rays obtained with a Ge(Li) high resolution detector (Debertin and Helmer, 1988, Figure 3.3). (b) Iodine KX ray escape from the 140 keV peak of a ^{99m}Tc spectrum obtained with a NaI(Tl) detector (Crouthamel, 1960, Appendix II).

the high-resolution spectrum of yttrium-88 (Figure 6.6(a)). The full energy peaks of these spectra are their outstanding characteristics. The detectors are calibrated by plotting the full energy peak efficiency of the γ rays (ε_γ) as functions of the γ ray energy (E_γ). Figure 6.7(a) is an example of such a calibration graph.

Before turning our attention to full energy peak efficiency in Section 6.4.5, we must discuss the following: pair production leading to escape peaks, the energy calibration of detectors and the role of energy resolution. Detailed information on recent developments is available from suppliers of these instruments (Section 4.3.3).

6.4.2 Escape peaks

Although NaI(Tl) and semiconductor detectors differ greatly in energy resolution (Figure 3.10), the spectra are generated by the same γ ray interactions: photoelectric, Compton and pair production (Figure 3.7(a)).

Escape peaks appear in γ ray spectra when high-energy γ rays ($E_\gamma >$ 1022 keV) interact in the detector by pair production, which has to involve an atomic nucleus to conserve momentum. The positron of the pair promptly undergoes further interactions which lead to escape peaks (Figure 6.6(a)).

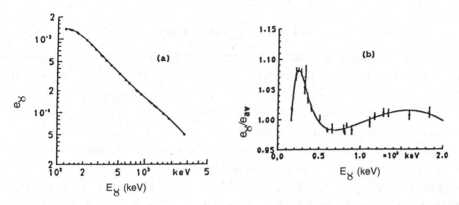

Figure 6.7. (a) A log–log plot of detection efficiency ε_γ versus γ ray energy E_γ(keV) for a high-purity germanium detector in the range 100 to 1500 keV. (b) Fractional differences ($\varepsilon_\gamma/\varepsilon_{av}$) from the straight line approximation (Debertin and Helmer, 1988, Section 4.2).

Pair production remains unlikely until γ ray energies exceed about 2 MeV (Figure 3.7(a)). When it occurs, the γ ray is transmuted into a negatron–positron pair as shown in Figure 4.1 (γ_3). This figure also shows the positron being annihilated in turn by another negatron (negatrons are ubiquitous in matter), leading to the creation of two annihilation γ rays each of energy 511 keV, leaving the point of origin in opposite directions.

If both γ rays are absorbed in the detector, the result is a pulse in the full energy peak. If one or both escape, the resulting pulses will be at energies $E_\gamma - 511$ keV and $E_\gamma - 1022$ keV (Figure 6.6(a)). These energy differences are readily verified and should confirm the energy calibration of the detector. Pair production due to γ rays of energy below about 1200 keV is of very low intensity and so difficult to identify.

When working with NaI(Tl) detectors, escape peaks also occur at the low-energy end of the spectrum due to the escape of averagely 29.2 keV fluorescent iodine KX rays. Iodine escape is triggered by sufficiently low-energy X or γ rays (but they must exceed 30 keV), interacting with iodine atoms in the NaI(Tl) detector when a γ ray of energy E keV is left with $(E-29.2)$ keV (Figures 3.1 and 6.6(b)), forming an escape peak at that energy. For escape to occur the KX rays must have been triggered sufficiently close to the surface of the detector. Iodine escape peaks are no longer identifiable when the full energy peak exceeds about 160 keV. In that case, the escape peak is absorbed into the full energy peak.

6.4.3 Energy calibrations

Gamma ray spectrometry serves to identify the radionuclide which emitted the rays and to measure its activity. It is then also necessary to identify the energies of all γ rays of interest. This requires an energy calibration of the detector to permit measurement of detected γ ray energies which is usually done as a function of their pulse heights.

Since pulse heights due to semiconducting detectors are linear functions of the γ ray energy, the energy calibration of these detectors is relatively straightforward. For a 2000 channel MCA correctly zeroed and using the linear amplifier to set the 1408 keV peak from a ^{152}Eu source (Figure 3.14(a)) into channel 1408, the 121.8 keV peak can be expected to be within ±1 channel of channel 122 (Debertin and Helmer, 1988, Section 3.4.2).

Pulse height peaks due to NaI(Tl) detectors are not as linear a function of the γ ray energy as those of germanium detectors. Also, they are less well defined. This is evident in Figures 3.9 and 3.10. Energy calibrations have to be effected using radionuclides emitting γ rays at single or well separated energies. As a rule the intensities of pulse height peaks vary more than linearly when γ ray energies are below about 300 keV but less than linearly at higher energies. For collections of γ ray energies and intensities readers are referred to the references listed in the last paragraph of Section 5.5.1.

6.4.4 Energy resolution

A high energy resolution is one of the most important characteristics of γ ray detectors. The superiority of the resolution of semiconducting detectors over NaI(Tl) detectors is illustrated in Figure 3.10.

Unless distorted by scatter or absorption effects, pulse height peaks due to good-quality γ ray detectors can be fitted to Gaussian peaks, as will be shown. The fit will be the closer the nearer the ratio FWTM/FWHM is to the Gaussian value of 1.82 when FWHM/σ_G can be expected to equal 2.36. Here FWTM stands for the full width of the peak at one tenth of its maximum height and σ_G the Gaussian standard deviation (Eq. (6.12)).

FWHM values (w keV) of full energy γ ray peaks have to be measured for each detector and each energy. They are proportional to the square root of the energy of the γ rays (E_γ) and so are the greater the greater the γ ray energy. However, they decrease when expressed in fractional terms (w/E_γ), so that energy resolution improves with increasing γ ray energies. However, it is possible to make usefully accurate estimates from the following approximations listed by manufacturers of γ ray detectors in their catalogues:

For NaI(Tl) detectors $\qquad\qquad w \approx 1.3 \times (E_\gamma)^{1/2}$ keV $\qquad$ (6.2)

For Ge(Li) or HPGe detectors $\quad w \approx 0.05 \times (E_\gamma)^{1/2}$ keV. $\qquad$ (6.3)

6.4.5 *Full energy peak efficiency calibration*

Introduction

Although the room temperature density of NaI is much lower than that of germanium (3.67 g/cm^3 as against 5.46 g/cm^3 for Ge), NaI(Tl) detectors are offered with higher detection efficiencies because they can be made in large sizes (Figure 4.4) at acceptable costs. Semiconducting compound crystals have lower detection efficiencies still because of their very small size (Figure 5.6(b)). (Readers are reminded that there are many other inorganic scintillators besides NaI(Tl), as pointed out in Section 4.3.3.)

Gamma ray spectrometry requires a calibration of the full energy peak efficiency of the detector as a function of the γ ray energy (Figure 6.7). Calibrations yield optimum results when made with radionuclides which emit each only a single γ ray of well known energy and intensity or γ rays forming clearly separated peaks, though peaks are much more readily separated using germanium detectors (Figure 3.10).

Preparatory procedures

The first step is to decide on the energy range for the calibrations and purchase the radioactivity standard sources needed to cover this range (Table 8.1), noting the energies of the γ rays of interest and also their γ ray intensities (f_γ). It is helpful to enter these data into a table similar to Table 4.1.

There are three other points to consider. (a) The accuracy of calibrations (and of subsequent measurements) depends critically on a strictly reproducible source–detector geometry, commonly achieved with a source stand as shown in Figure 6.8. These stands are made of thin but rigid plastic rods to minimise γ ray scatter. (b) Detector and source should be shielded from unwanted radiations by 5 cm thick lead bricks. (c) When effecting calibrations it is always essential to estimate and state all uncertainties.

The calibration

Calibrations to determine full energy peak efficiencies (ε_γ) of a high-resolution detector as a function of the γ ray energy E_γ are discussed in detail by Debertin and Helmer (1988, Section 4.2.1). Nor surprisingly, calibrations are most accurate when working with radionuclides that emit only a single γ ray

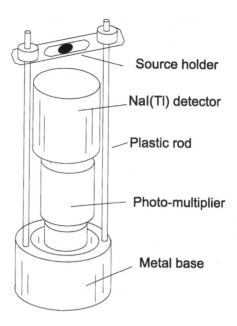

Figure 6.8. An adjustable source stand made from thin plastic rods to ensure reproducible source–detector geometry for γ ray spectrometry while minimising γ ray scatter.

of well known intensity (f_γ). When working with semiconducting detectors one could use multi gamma ray emitters such as ^{152}Eu and ^{133}Ba (Section 3.7.3) though there could be significant uncertainties in individual f_γ values.

Let us assume a source of ^{152}Eu of activity N_0 Bq and a high-resolution spectrometer. The latter is set up as required, corrections are measured and so is the reference time (Section 6.5.1). Let the calibration begin with the 1408.0 keV peak, the highest energy peak which is clearly displayed (Figure 3.14(a)) where $f_\gamma = 20.8\%$, so that N_γ, the γ ray emission rate from the source, is equal to 0.208 N_0. The ε_γ–f_γ relationship is then given by:

$$\varepsilon_\gamma = (N_p/N_\gamma) = N_p/(f_\gamma \, N_0). \tag{6.4}$$

$(N_p)_{1408}$, the countrate in the 1408 keV full energy peak, is obtained using a computerised peak integration system built into the MCA (Section 6.4.1). One could proceed similarly for the other ^{152}Eu peaks, to obtain a sufficient number of results for ε_γ and f_γ and their logarithms to draw the calibration graph which will be similar to the graph shown in Figure 6.7.

With the equipment in exactly the same source–detector geometry as for the calibration, one can use the graph, for example, to measure activities produced by neutron activation analysis (NAA) (Section 7.4.4).

Let it be assumed, for simplicity, that the γ ray spectrum generated by the NAA contains only a single peak, N_p' of energy E_γ. N_p' cps is measured by the MCA (see above) when the corresponding value of ε_γ is read from the calibration graph. With these values known, it is commonly straightforward to identify the emitting radionuclide and the relevant f_γ value from a table of nuclides, e.g. *Nuclides and Isotopes* (1989) or CoN (1998). For internet access there is also Chu *et al.* (1999) but here again, experimenters could find themselves with much more data than would be of interest to them. When the radionuclide data are known, N_0 Bq, the activity of the identified nuclide, is obtained by re-arranging Eq. (6.4) to give:

$$N_0 = [N_p'/(f_\gamma \times \varepsilon_\gamma)]. \tag{6.5}$$

When a detector has been calibrated, activity measurements can be carried out on any peak of a γ ray emitter for which f_γ is known with sufficient accuracy, provided the background rate and other corrections are properly allowed for. If this is done for all peaks due to a multi gamma ray emitter, the agreement between N_0 values calculated from the countrates of each peak of the same radionuclide (assumed large enough for accurate measurements), can be a useful indicator of the soundness of the relevant decay data and/or the calibration.

As shown in Figures 6.7(a) and (b), efficiency calibrations of germanium detectors have the useful characteristic that a plot of log ε_γ versus log E_γ for a selected source–detector geometry and over the energy range about 120 to 1500 keV is a straight line (to a close approximation). Efficiency calibrations could be based on four evenly spaced peaks from a single multi γ ray emitting radionuclide, but not at high accuracy.

6.4.6   Secondary standard instruments: strong and weak points

Several of the strong points in favour of pressurised ionisation chambers are relatively weak in γ ray spectrometers and vice versa. Laboratories requiring reliable radioactivity measurements should have access to both types of instrument.

Pressurised ionisation chambers are highly stable instruments, principally on account of their simple construction (Figure 6.2). Also, given their large volume (Table 6.2), the electric field near their centre is sufficiently uniform to permit a relatively large displacement of small-volume sources with little change in the ionisation current (Figure 6.4(b)). In contrast, the source–detector geometry of γ ray spectrometers has to be carefully monitored because any changes are amplified by the inverse square law.

Well shielded γ ray detectors can measure countrates down to a single cps above background with only a few per cent uncertainty. In contrast, pressurised ionisation chambers cannot identify differences in ionisation currents smaller than those due to about 1 kcps of γ rays. The lower limit depends on the intensities and energies of the emitted radiations, being up to 80 times lower for γ rays from ^{60}Co than for γ rays from equal activities of ^{51}Cr.

As their strongest points, γ ray spectrometers serve (a) to identify γ ray emitting radionuclides (by identifying the energies of full energy peaks) and (b) to detect γ rays due to small concentrations of radionuclidic impurities or radiotracers, and this with high sensitivity and for a large number of radionuclides. Pressurised ionisation chambers are unsuitable for such tasks.

The lower γ ray sensitivity of the gaseous detection media in ionisation chambers is compensated by their 4π detection geometry and large volume. Also, the calibration graph of pressurised ionisation chambers slopes upwards (Figure 6.5) as distinct from the downward slope for γ ray detectors (Figure 6.7). As γ ray energies increase, so do the energies of the electrons knocked by the γ rays into the gaseous detection medium, thus increasing the ionisation current.

6.5 Results, part 1: collecting the data

6.5.1 Five components for a complete result

Many results of measurements of countrates or of ionisation currents made in applications are only provisional when it could be sufficient to simply quote a countrate or an ionisation current and leave it at that. For a result to be meaningful to others it should include five components as follows.

(i)   The mean of the series of measurements made for the application of interest.
(ii) The date and time of day to which the mean refers, called the reference time, T_r. This is an essential a part of the result as, for example, the countrate.
(iii) An uncertainty statement – results are invariably subject to uncertainties that must be estimated and stated.
(iv) The confidence limit stating the level of confidence of the experimenter that the estimated uncertainty will not be exceeded.
(v)   A statement about the radionuclidic purity of the radioactive material.

At this point only components (iii) and (iv) still need explanatory comments. The following will offer a few guidelines on how errors and uncertainties can be identified and evaluated.

6.5.2 *Errors and uncertainties*

Results of physical measurements are likely to be subject to errors, e.g. failure to correct for the background. They are also subject to uncertainties, e.g. uncertainty about nuclear data obtained from the literature and uncertainties in necessary corrections. Results of measurements of radioactivity and many other properties are subject to two types of uncertainties known as random and systematic uncertainties or also as type A and type B uncertainties. While errors could be avoided at least in principle and this will be here assumed, uncertainties cannot be avoided

Random uncertainties apply for instance to measured countrates due to random fluctuations in the rate of radioactive decay (Section 2.2.1). Statistical methods described below are used to estimate these uncertainties. On the other hand, systematic errors arise e.g. from uncertainties in the radionuclide decay rates, the gamma ray fraction, the presence of radionuclide impurities or instrumental problems. Estimates of systematic errors are, at least in part, subjective, bearing in mind that the parameters have been measured elsewhere.

Table 6.3 shows a list of possible random and systematic uncertainties. They were calculated or estimated following a calorimetry measurement of the activity of a pure β particle emitter as described in Section 5.4.3 and Figure 5.5 (see Genka *et al.*, 1987). It is not uncommon for results of radioactivity measurements to be affected by four or more uncertainties that have to be combined by realistic and reliable methods to arrive at the overall uncertainty, which should be quoted with the final result.

To calculate the overall uncertainty, say U, in the mean of a series of measurements, e.g. of countrates, systematic uncertainties (U_s) are added linearly regardless of sign i.e. $U_s = U_{1s} + U_{2s} + \cdots U_{ns}$ whereas random uncertainties (U_r) can be added in quadrature, $U_r = (U_{1r}^2 + U_{2r}^2 + \cdots U_{nr}^2)^{1/2}$. It is readily verified that linear additions add more weight to the overall uncertainty than quadrature additions. For instance, assuming five uncertainty estimates, each $\pm 2\%$, the linear sum is 10% whereas the quadrature sum is only $(5 \times 2^2)^{1/2} \approx \pm 4.5\%$.

Quadrature summing is justified when corrections to a stated result could be as likely to increase an uncertainty as to decrease it. Although systematic uncertainties should be added linearly (see above), there are normally several of them whence it is often realistic to assume that the overall effect is as likely to decrease the total uncertainty as to increase it, so justifying quadrature summing to obtain overall systematic uncertainties. Treatments of statistical uncertainties are given by Bevington (1969), Kirkup (1994, Ch. 4), Spiegel

Table 6.3. *Corrections and uncertainties applying to a calorimetry measurement of the activity of phosphorus-32.*[a]

Corrections (%)	
Heat loss due to thermal radiation	+2.2
Bremsstrahlung escape	+0.1
Radionuclidic impurity	−0.1
Uncertainties (±2σ)[b]:	
Random:	
Thermal power measurement	±0.90
Systematic:	
Average energy of beta spectrum	±0.13
Uncertainty in half life	±0.11
Calibration of the calorimeter	±0.50
Correction for heat loss[c]	±0.30
Correction for bremsstrahlung escape	±0.05
Correction for radionuclidic impurity	±0.02
Overall uncertainty (quadrature sum)	±1.09

[a] See notes in Table 5.2 for references to this work.

[b] ±2σ denotes the confidence limits for which the stated uncertainties were quoted (see Table 6.4(a)).

[c] Note that each correction has its associated uncertainty.

(1999) and by the US National Institute of Science and Technology (NIST) via the Internet (see Appendix 3).

6.6 Results, part 2: Poisson and Gaussian statistics

6.6.1 A first look at statistical distributions

Systematic uncertainties, e.g. in a half life, have to be estimated as best one can, being guided by conventions and the uncertainty estimates of the experimenters who measured the half life. In contrast, uncertainties in randomly occurring variables can be expected to follow statistical distributions when they can be estimated or verified using statistical theories. It is assumed here that measured countrates are proportional to the randomly occurring nuclear decay rates which gave rise to them.

Randomness could be disturbed by malfunctioning equipment or other external factors or by errors in a procedure. One should always make sure that it applies within acceptable uncertainties. The counting equipment should be monitored by frequent measurements of a high-quality, long-lived radioactive source, preferably radium-226 in equilibrium with its daughters

(Table 6.1). Several tests have been designed to verify that observed countrates are randomly distributed as well as can be expected and a few of these tests will be introduced in Sections 6.6.4 and 6.7.

The Gaussian distribution is symmetrical, continuous and easily applicable to numerous very different situations throughout the natural and social sciences. It can be fitted to radioactive decay rates but only if countrates exceed about 15 cps (see below). The Poisson distribution was invented to take care of asymmetrical situations where there are large numbers of probable outcomes but only a small fraction is actually realised. Radioactive decays occur in conditions where there is a large number of unstable atoms (N), but the numbers which decay in unit time ($-dN/dt$) are only a small or very small fraction of N; this is the Poisson criterion. Using Eqs. (1.5) and (1.6) from Section 1.6.2 it can be shown that:

$$N/(dN/dt) = T_{1/2}/\ln 2. \tag{6.6}$$

With dN/dt in dps, $T_{1/2}$ has to be stated in seconds and $\ln 2$ equals 0.693. For normally used radionuclides one rarely has $T_{1/2}$ less than several hundred seconds, giving $T_{1/2}/\ln 2$ of the order of 10^2, i.e. sufficiently large to satisfy the Poisson criterion. For ^{60}Co one has $T_{1/2} = 5.27$ years when the ratio is of order 10^8.

More detailed information on the role of statistics for radioactivity measurements can be found in Debertin and Helmer (1988, Section 1.5), in specialised texts (e.g. Bevington and Robinson, 1992) or via Internet sites (see Table A3.1 in Appendix 3).

6.6.2 The Poisson distribution

The Poisson distribution is generated from the expression:

$$P(n) = m^n \times e^{-m}/n! \tag{6.7}$$

where $P(n)$ is the normalised probability ($\sum[P(n)] = 1$)for successive values of $n(\geq 0)$ and is defined only for positive integers and for zero; m (in the exponent) is the mean of the distribution, commonly the arithmetic mean obtained from as many repeat counts as possible, with all results equally probable or weighted to allow for inequalities. It is also assumed that countrates have been fully corrected as required in the circumstances.

Observed decay rates are either zero or positive integers, causing countrate distributions to be asymmetric about m, as is the Poisson distribution. This is clearly seen for low countrates (Figures 6.9 and 6.10). However, at $m = 20$ the asymmetry has become unobservably small (Figure 6.9) which is readily

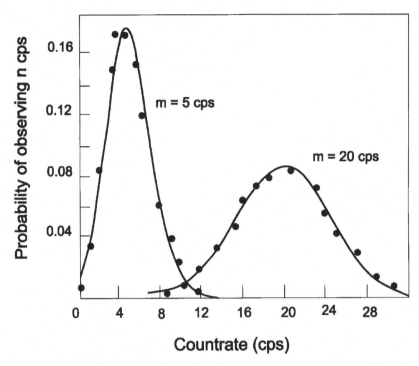

Figure 6.9. Poisson (dots) and Gaussian (continuous curves) probability distributions calculated from Eqs. (6.7) and (6.10).

verified from Eq. (6.7). If m is less than 10 cps (or certainly if it is less than 5 cps) results should be processed by Poisson statistics which allow for the asymmetry in the data.

When Poisson statistics apply, the variance, v (which is a measure of the spread of the distribution about m) is numerically equal to m which, in turn, is equal to the square of the standard deviation σ_P. Taking square roots one obtains:

$$m^{1/2} = v^{1/2} = \sigma_P. \tag{6.8}$$

The fractional standard deviation σ/m is another useful measure of the accuracy of results. It is defined as:

$$\sigma_P/m = m^{1/2}/m = m^{-1/2} = 100m^{-1/2}\%. \tag{6.9}$$

It follows from Eq. (6.9) that it is necessary to accumulate at least 10^4 counts per measurement ($m > 10^4$) to ensure that σ/m is better than 0.01 or 1%.

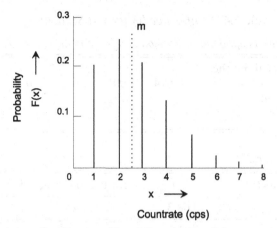

Figure 6.10. The Poisson distribution (Eq. 6.7) calculated for a mean (*m*) of 2.5 cps when the asymmetry is pronounced (Debertin and Helmer, 1988, Figure 1.22).

In this chapter calculations will be concerned only with *m* and σ, which should always be calculated from as many repeat measurements as practical, all taken in identical conditions. There are many applications when it is impractical to make more than a few repeat measurements in identical conditions, so adding to uncertainties which can be allowed for using the Student *t* factor (Table 6.4(b)).

6.6.3 Gaussian statistics

The Gaussian distribution deals with differences between results of individual counts n_1, n_2, . . . , n_i (all taken in identical conditions and corrected as required), and their arithmetic mean *m*. Being symetrical, it can be applied, e.g. to the scatter of results of repeated weighings of a solution about their mean. These and similar results follow Gauss statistics but not Poisson statistics.

Writing $m - n_i = \delta_i$, the normalised Gauss function is generated from the expression:

$$P(n)_G = w \exp[-(\delta^2 I)/(2\sigma_G^2)], \tag{6.10}$$

where *w* is a constant depending on the Gaussian standard deviation σ_G and so on the width of the distribution which, when processing decay rates, has to conform to the width calculated in accordance with the Poisson distribution.

As noted earlier, both *m* and σ_G should be calculated from as many repeat measurements as possible, all taken in identical conditions, so obtaining n_i results (degrees of freedom). For the large majority of applications *i* would

Table 6.4. *Confidence limits and Student t factors.*

(a) Confidence limits applying to Poisson and Normal distributions (Section 6.6.4).

Confidence limits about the mean as multiples of the standard deviation σ	0.674	1.00	1.64	2.00	3.00	3.30
Probability (%)[a]	50	68.3	90	95.4	99.7	99.9

(b) Student t factors.[b]

Number of counts (n)	t factors for use in the stated intervals, see (a)		
	±68.3	±95.4	±99.7
3	1.32	4.3	19.2
4	1.20	3.2	9.2
5	1.15	2.8	6.9
6	1.11	2.6	5.5
8	1.08	2.4	4.5
10	1.06	2.3	4.1
20	1.03	2.1	3.4
50	1.01	2.0	3.2
100	1.00	2.0	3.1

[a] These are the probabilities that countrates will, on average, fall within the stated range.

[b] If five successive measurements of a countrate yield $m = 25.0$ cps the standard deviation is not simply $25.0^{1/2}$ but $\pm 1.1 \times 25.0^{1/2}$, i.e. ± 6 to the nearest integer. 25 ± 5 could be assumed to apply for >20 repeat measurements. t factors are often ignored; whether or not they should be used depends on the required accuracy.

rarely be as large as 20. Uncertainties due to a smaller number of repeat measurements could be allowed for using the Student t factor (Table 20, NCRP, 1985). Bearing this in mind, the arithmetic mean is given by:

$$m = \sum_{i=1}^{i} n_i / i. \tag{6.11}$$

Having to calculate m reduces the number of independently obtained results (degrees of freedom) from n_i to $(n_i - 1)$.

One can now calculate σ_G from the sum of the squares of the differences between the mean m and individual n_i values i.e. $\sum_{i=1}^{i}(m - n_i)^2$ when:

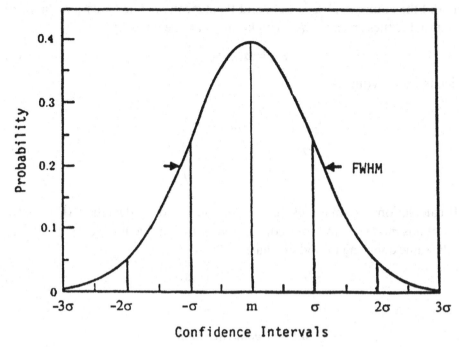

Figure 6.11. A Gaussian distribution calculated from Eq. (6.10), showing confidence intervals in terms of the standard deviation (Eq. (6.12)).

$$\sigma_G = \left(\frac{\sum\limits_{i=1}^{i}(m - n_i)^2}{(n_i - 1)} \right)^{1/2}. \tag{6.12}$$

With the mean m, σ_P and σ_G defined and assuming that m is large enough for Gaussian statistics to apply, it can be expected (Table 6.4(a)), that 68.3% of the individual results are, on average, within $(m \pm \sigma_G)$ and 31.7% are outside these limits. It should be rare (on average no more than one case in twenty) that the difference exceeds $\pm 2\sigma_G$, and in only about three cases in 1000 should the difference exceed $\pm 3\sigma_G$ (Figure 6.11). If differences between the mean and individual counts are significantly smaller or larger than expected from the rules just stated, this should be investigated in case randomness has been interfered with.

The series of measurements used to calculate m in Eq. (6.11) could be repeated, say n_k times in otherwise identical conditions to obtain a series of means, $m_1, m_2 \ldots, m_k$. Making decay corrections, if necessary, and assuming randomness, these mean values will agree more closely among themselves than did the counts $n_1, n_2, \ldots, n_i$ and their mean m. The Gaussian standard deviation for these means, σ_{Gm}, can be shown to be given by

$$\sigma_{Gm} = \sigma_G/(n_k)^{1/2}, \tag{6.13}$$

which is closely equal to:

$$\sigma_{Gm} = \left(\frac{\sum\limits_{i=1}^{i}(m - n_i)^2}{n_k(n_i - 1)} \right)^{1/2}. \tag{6.14}$$

If calculations are made using the Poisson standard deviation (Eq. (6.8)), one obtains $\sigma_P = m^{1/2}$ and for i consecutive counts in identical conditions and for the same counting period one has:

$$\sigma_{Pm} = (i \times m_i)^{1/2} = 100(i \times m_i)^{-1/2}\%. \tag{6.15}$$

6.6.4 Confidence limits

Confidence limits provide information on the reliability of uncertainty statements and should always be quoted together with the uncertainty to which they refer. When countrates conform to Poisson and Gaussian statistics, the confidence limit is equal to the standard deviation or a multiple thereof (Table 6.4(a)). However, occasional deviations from a result expected on statistical grounds could always occur by chance.

If the number of repeat counts has to be kept small (say <10), the verification of randomness should be effected using the Student t factor (Table 6.4(b)). For instance, to be confident of a 68.3% confidence level for a distribution due to only six repeat counts, the calculated value of σ_P (Eq. (6.9)) should be multiplied by 1.11 (Table 6.4(b)). To realise a 99.7% confidence level, the multiplier is not 3 (Table 6.4(a)) but 5.5 (Table 6.4(b)). This increase in the standard deviation is introduced to compensate for the added uncertainty due to the small number of repeat measurements.

The situation is different when estimating confidence levels to allow for systematic uncertainties (Section 6.5.2). To obtain reliable estimates it may be advisable for the measurements to be repeated often enough to permit this to be done realistically since confidence levels, to be helpful, should be neither too large nor too small.

6.7 Other characteristics of results and statistical tests

6.7.1 Countrates and their combination

Whenever practical, results of counts of radioactive decays should be stated as countrates, $r = n/t$ where n is the number of counts in time t seconds so that n/t is expressed in counts per second (cps).

If the mean of a series of countrates is r_m, Poisson statistics predict:

$$r_m \pm \sigma_m = n/t \pm n^{1/2}/t = r_m \pm (r_m/t)^{1/2}, \qquad (6.16)$$

while the fractional standard deviation (Eq. (6.9)), expressed as a countrate becomes:

$$
\begin{aligned}
r_m \pm (r_m/t)^{1/2}/r_m &= r_m \pm 1/(r_m t)^{1/2} \\
&= r_m \pm 100/(r_m t)^{1/2}\% \qquad (6.17) \\
&= r_m \pm 100/n^{1/2}\%
\end{aligned}
$$

since $r \times t = n$ (see above).

There are experimental situations when several results, A, B, . . . , each with its standard deviation, σ_A, σ_B, . . . , have to be combined by either addition or subtraction. One obtains the standard deviation σ_S of the sum or difference by quadrature summing (Section 6.5.2), i.e.

$$\sigma_S = (\sigma_A^2 + \sigma_B^2 + \cdots)^{1/2}. \qquad (6.18)$$

Similarly, when two or more results are combined as a product, say $R_1 \times R_2$ or a quotient R_1/R_2 with standard deviations σ_1 and σ_2, the fractional standard deviation of the product σ_P/P is given by:

$$\sigma_P/P = [(\sigma_1/R_1)^2 + (\sigma^{1/2}/R_2)^2]^{1/2}. \qquad (6.19)$$

Results of statistical distributions are the more accurately defined the larger their mean m and therefore the smaller the fractional standard deviation σ/m. However, all countrates must be equally probable or they should be weighted to allow for any differences.

If countrates are low, a large value of m can be realised only over a long counting time. This could create difficulties due to instabilities in the counting equipment especially so if the count plus background rate, r_{sb} is similar to the background rate, r_b. It is then helpful to divide the available counting time between t_b and t_{sb} so that r_s, the countrate due to the source calculated from $r_{sb} - r_b$ is measured with the smallest realisable uncertainty. It can be shown that this applies if the counting times t_{sb} and t_b per source are in the ratio:

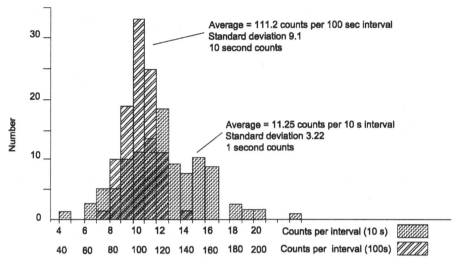

Figure 6.12. Two histograms each obtained for 100 repeat counts at a count rate close to 1.1 cps. (a) The histogram for 10 s counts. Average = 11.25, standard deviation = 3.22. (b) The histogram for 100 s counts. Average = 111.2, standard deviation = 9.10 (Shapiro, 1972, Fig. 4.19).

$$t_{sb}/t_b = (r_{sb}/r_b)^{1/2}. \qquad (6.20)$$

The standard deviation applicable to the source is then:

$$\sigma_s = (r_{sb}/t_{sb} + r_b/t_b)^{1/2}. \qquad (6.21)$$

Clearly set out examples are offered by Shapiro (1972, Pt. IV, Section 6).

6.7.2 Tests for accuracy and consistency

Accuracy

Figures 6.12(a) and (b) show two superimposed histograms of counting results. Each was obtained with 100 repeat counts taken in identical conditions at a countrate close to 1.1 cps, but the counting times were respectively 10 and 100 seconds. The standard deviation, σ_G, was calculated from Eq. (6.12). For the 10s counts, σ_G/m was 3.22/11.25 i.e. σ_G is about 29% of m; for the 100s counts σ_G/m equals 9.10/111.2 i.e. σ_G is about 8.2% of m. The result is a significantly more tidy histogram. For highly accurate results one aims at a near-smooth distribution when σ/m values should be 0.1% or less. In that case a plot of the observed data (assuming random conditions), can be expected to closely approximate a Gaussian curve. The ratio of the two σ_G/m values, 29%/8.2%, is 3.5, which is close to the ratio of the square root of the mean counts, $111.2^{1/2}/11.25^{1/2}$, or 3.1 as predicted by Eq. (6.9).

Figure 6.11 shows a Gaussian distribution obtained from a very large number of calculated results (Eq. (6.10)) so explaining the smooth curve in contrast to the histograms in Figure 6.12. The area between the curve and the base line (the Gaussian function extends from $-\infty$ to $+\infty$) represents all possible results while the areas between $m \pm 1\sigma$, $m \pm 2\sigma$ and $m \pm 3\sigma$ represent respectively 68.3, 95.4 and 99.7% of the total to a high degree of accuracy (Table 6.4(a)).

Consistency

It is often important to verify that conditions in a system that is regularly employed for countrate measurements either remain unchanged from one day to the next (or longer), or change by a specified amount in line with changes in experimental parameters. A procedure developed for such tests, which causes minimum interference with normal requirements, is the so-called t-distribution. The data used for the example below can be found in NCRP (1985, Table 20).

Following two sets of measurements taken at different times but otherwise in identical conditions, one calculates the t factor from the expression:

$$t = (m_i - m_2)/[(\sigma_1^2 + \sigma_2^2)^{1/2}/\delta n]^{1/2}, \qquad (6.22)$$

where the symbols have the same meaning as in Section 6.6.3 and σ could be obtained either from Eq. (6.8) or Eq. (6.12). For a comparison between two sets of readings one has $\delta n = 2$.

Let sets of ten readings each be taken with a long-lived source (decay is negligible) one week apart, to verify the stability of a process. Suppose the recorded results were: $m_1 = 3830$, $m_2 = 3604$, $\sigma_1 = 87.8$, $\sigma_2 = 195.1$, with $\delta n = 2$ when Equation (6.22) then yields $t = 2.99$.

From Table 20 in NCRP (1985) the critical value for 18 degrees of freedom $(10 + 10 - 2)$ and a probability level of 0.01 is close to 2.90, less than the calculated 2.99, so suggesting that the probability for these two sets to indicate consistent conditions is less than 1% and certainly less than 5%, suggesting that the process could have been disturbed.

6.7.3 *Tests for randomness*

A simple test to verify randomness is known as the double check test. It can be applied when the mean, m, exceeds at least 15 and so is large enough for the standard deviations to be calculated either by Poisson or Gaussian

Table 6.5. *A selection of chi squared values.*

Degrees of freedom $(n-1)$[a]	There is a probability of						
	0.99	0.95	0.90	0.50	0.10	0.05	0.01
	that the calculated value of chi squared will be equal to or greater than:						
5	0.554	1.145	1.610	4.351	9.236	11.070	15.086
6	0.872	1.635	2.204	5.348	10.645	12.592	16.812
7	1.239	2.167	2.833	6.346	12.017	14.067	18.475
8	1.646	2.733	3.490	7.344	13.362	15.507	20.090
9	2.088	3.325	4.168	8.343	14.684	16.919	21.666
10	2.558	3.940	4.865	9.342	15.987	18.307	23.209
11	3.053	4.575	5.578	10.341	17.275	19.675	24.725
12	2.571	5.226	6.304	11.340	18.549	21.026	26.217
13	4.107	5.892	7.042	12.340	19.812	22.362	27.688
14	4.660	6.571	7.790	13.339	21.064	23.685	29.141
15	5.229	7.261	8.547	14.339	22.307	24.996	30.578
16	5.812	7.962	9.312	15.338	23.542	26.296	32.000
17	6.408	8.672	10.085	16.338	24.769	27.587	33.409
18	7.015	9.390	10.865	17.338	25.989	28.869	34.805
19	7.633	10.117	11.651	18.338	27.204	30.144	36.191
20	8.260	10.851	12.443	19.337	28.412	31.410	37.566
21	8.897	11.591	13.240	20.337	29.615	32.671	38.932
22	9.542	12.338	14.041	21.337	30.813	33.924	40.289
23	10.196	13.091	14.848	22.337	32.007	35.172	41.638
24	10.856	13.848	15.659	23.337	33.196	36.415	42.980

[a] The number of degrees of freedom is usually one less than n, the number of repeat measurements.

statistics. If the results obtained with the two statistics agree within less than about 10% this can be accepted as evidence for randomness. A failure to pass this test, especially when values of m are known to be accurate, say ±1% or better, should put experimenters on the alert when looking for errors.

A more powerful test is the χ^2 (or chi squared) test. The data needed for this test when the number n of repeat counts is between 6 and 25 are shown in Table 6.5. Using the same notation as in Equation (6.12), the χ^2 function can be shown to be defined by:

$$\chi^2 = \sum_{i=1}^{i}(m - n_i)^2 \bigg/ m. \tag{6.23}$$

For a distribution which obeys Poisson and Gaussian statistics, one has:

$$\sigma^2 = m$$

(see Eq. (6.8)). Also squaring both sides of Eq. (6.12) yields:

$$\sigma_G{}^2 = \sum_{i=1}^{i} (m - n_i)^2 / (n_i = 1).$$

Substituting these relations into Eq. (6.23) yields:

$$\chi^2 \approx (n_i - 1). \tag{6.24}$$

Here n_i again represents the number of equally probable observations (see Table 6.5). One degree of freedom was used to calculate the mean, making it necessary to replace n_i by $n_i - 1$ as was done in Eq. (6.12).

A high probability for randomness applies when χ^2 does not differ greatly from 0.50 (50%) though many experimenters accept χ^2 in the range 10 to 90% as admissable evidence for randomness. However, results smaller than 10% or larger than 90% strongly suggest non-randomness.

Suppose an experimenter measured the countrate from a radioactive source, accumulating 20 counts to obtain $m = 312$ cps, $\sigma_G = 5.3$ cps, $\sigma_P = 3.9$ cps and $\chi^2 = 33$. These results suggest a significant non-random effect. The double check test shows a difference between σ_G and σ_P of over 30% instead of under 10% as expected for randomness. Also, with $(n - 1) = 19$ (Table 6.5), a χ^2 value of 33 suggests a less than 5% probability for randomness. Given these very different results from what is expected for randomness, the experimenter has to discover what caused them and how the results could be improved.

6.8 Moving on to applications

This chapter ends the introductory part of this book. It was written to familiarise readers, if only to some extent, with the characteristics of the nuclear radiations commonly employed for applications in industry, technology and related fields, and with the basic properties of the radionuclides that emit these radiations. This was done at somewhat greater length than is normal for books dealing with applications because the basic nuclear sciences are no longer as widely taught in schools and universities as they used to be. Nevertheless, experimenters will frequently find it necessary to look for additional information among the quoted references, in other textbooks or via the Internet, as listed in Table A3.1 in Appendix 3.

The next three succeeding chapters are the core of this book. The nuclear radiation applications discussed in these chapters will often deal with large-scale undertakings: malfunction in large industrial plants, extensive searches for water or oil, or safe procedures for the dispersal of high-activity nuclear waste to name just a few. Descriptions of large projects make it necessary to arrange the material in Chapters 7, 8 and 9 in a somewhat different way from what went before. Whenever practical, use will be made of the foregoing discussions of the characteristics of radionuclides and nuclear radiations. More often it will be necessary to refer experimenters to the scientific literature. On the whole readers will be encouraged to develop their own procedures and apply them to suit their interests.

Chapter 7

Industrial applications of radioisotopes and radiation

7.1 Introduction

7.1.1 A change of emphasis

Beginning with this chapter, the emphasis of this book will shift from discussing the science and technology of radioactivity and ionising radiations to considering how they are employed to advance scientific, technological and social objectives. The present chapter will concentrate on industrial applications of X and γ rays, charged particles and neutrons, and also on the applications of high-intensity radiations for chemical processing. Chapter 8 will discuss industrial radiotracing and Chapter 9 the contribution of nuclear techniques to understanding and protecting the environment.

7.1.2 An overview of industrial applications

Summary

The goal of modern manufacturing is to produce quality products as economically as possible using processes designed to minimise adverse impacts on the environment. The processes should result in the maximum efficiency in the use of energy and materials and the minimum generation of waste products.

Nuclear techniques contribute to this goal when applied to:

- optimising the efficiency of industrial processes;
- diagnosing problems in plant operations;
- examining industrial components using non-destructive methods;
- analysing the composition and structure of materials;

181

- treating materials with high-energy radiation for the purposes of sterilisation or modification of their properties;
- assessing the impact of industrial and urban development on the environment and designing remedial action.

The scope of these applications is outlined below.

Optimisation and control of processes in industrial plant

Optimisation and control of industrial processes depend first and foremost on the measurement of key parameters. In modern industrial plant, these data are often interpreted with the aid of sophisticated mathematical models designed to monitor the processes and to diagnose operational problems.

A wide range of nucleonic gauges has been developed for the acquisition of data on industrial plant. Examples include measurements of

- the levels and densities of materials in vessels and pipelines
- the thicknesses of sheets and coatings
- the amounts and properties of materials on conveyor belts.

All gauges comprise one or more radiation sources and a detector system optimised for the measurement of interest. In most cases the output is obtained in real time and can therefore be used not only for monitoring, but also for process control and optimisation. The gauges are typically mounted external to the plant and do not interfere with the process either during installation or operation.

Plant diagnostics

A range of radioactive tracer techniques has been designed to investigate reasons for any reduced efficiency in plant operation. For instance, tracers are routinely used to measure flow rates, study mixing processes and locate leaks in pipework and heat exchangers.

However, the role of radiotracers in plant diagnostics is fundamentally changing. The increasing use of mathematical models is reducing the need for tracers in many routine applications. On the other hand, radiotracers are being increasingly used for model validation and for the monitoring of processes which are so complex that detailed mathematical descriptions are inadequate.

Testing and inspection of materials

Long established industrial radiography is routinely used in studying the integrity of structural materials, especially the testing of welds. Recent advances in data processing have enabled the two or three dimensional visualisation of the internal structure of materials and components using a technique known as computerised tomography. Both radiography and tomography were first used in medicine but are now firmly established in industry.

Composition and structure of materials

Knowledge of the composition, structure and compatibility of materials is frequently required in the investigation phase of an industrial project. Neutron activation and X ray fluorescence techniques are widely used for elemental analyses, and a range of X ray, electron and neutron based methods have been developed for structure and compatibility studies.

Modification and syntheses of materials

The passage of sufficiently intense ionising radiation through matter leads to chemical and physical changes, many of which are important industrially. In this chapter reference will be made to the chemical synthesis of a range of polymeric materials by radiation induced reactions and to the widespread use of radiation for the sterilisation of medical products.

Environmental applications

Despite modern policies encouraging waste minimisation, it is inevitable that some unwanted by-products of development will be released to the environment. Applications of radioisotopes to the study of the transport of contaminants through terrestrial and marine ecosystems will be outlined in Chapter 9.

Broadly speaking, the topics are classified according to the radiation source. Emphasis is placed on the applications of gamma rays (Table 7.1) and beta and neutron radiation (see later – Tables 7.5 and 7.6 respectively). By organising the material in this way, reference can easily be made to the underlying scientific principles. Readers are referred to publications containing detailed information, for example the works of Charlton (1986), IAEA (1990a) and Hills (1999). As an indication of the relative importance of different applications, the distribution of commercial projects undertaken by a major service provider in Australia over the period 1989 to 1999 is listed in Table 7.2.

Table 7.1. *Industrial applications of gamma rays.*

Property of the Radiation	Application	Example	Reference
Attenuation	Level gauges	To monitor and control the levels of liquids and solids in chemical reactors, tanks and hoppers	
	Column scanning	Diagnosis of malfunction of industrial columns during plant operation	Section 7.2.1
	Density gauges	Density of materials in industrial and mineral processing streams	
	Industrial tomography	Visualisation of the internal structure of components and materials in 2 or 3 dimensions	
	Dual isotope	Ash in coal on conveyor belts	
Back-scatter	Level gauges	Levels of liquids in tanks	
	γ-γ lagging of boreholes	(a) Bulk density of strata (b) Monitoring of the water table (c) Monitoring of oil/water interface	Section 7.2.2
	On-line thickness monitoring	Monitor and control of surface coating thickness e.g. tin plate	
Fluorescence (X ray)	X ray Fluorescence (XRF)	Multi element assay	Section 7.2.3
Absorption	Ionisation chambers	Secondary standard instruments for radioactivity measurements	Sections 5.2, 6.3, 2.3
Ionisation (γ, X rays)	Gas proportional counters Geiger–Müller counters	Radiation detection and monitoring	Section 5.3

Table 7.1. *(cont.)*

Property of the Radiation	Application	Example	Reference
Chemical effects of high energy radiation	Chemical dosimetry; Radiation polymerisation; Radiation pasteurisation and sterilisation; Food irradiation	Industrial dosimetry; Good radiation practice; Radiation cross-linking – heat shrink plastics Sterilisation of medical products; Waste water treatment; Extend the post harvest life of certain foodstuffs.	Section 7.6.2 Section 7.6.3 Section 7.6.4 Section 7.6.5

Table 7.2. *Industrial applications of radioisotopes and their distribution in Australia, 1989–1999.*[a]

Category	Application	Per cent
Sealed source applications	Measurement of density profiles of distillation columns	19.3
	Detection of deposition and blockages in pipelines	9.6
	Measurement of levels and interfaces in process vessels	23.5
	Tomographic scans for density distribution measurement	1.8
	Flooded member detection (identification of water ingress into sub-sea structures)	5.8
	Measurement of corrosion/erosion in pipelines	1.7
	Neutron scans for detection of water under thermal insulation (corrosion pre-cursor)	1.5
Radioactive tracer applications	Measurement of residence time distributions in process vessels	11.7
	Flowrate measurements in liquids, gases, solids	7.4
	Fluid dynamics and mass transfer properties of fluidised catalytic crackers (FCCs)	5.2
	Internal leakage detection in process vessels	4.4
	Detection of underground leaks	1.7
	Liquid/gas distribution studies in chemical reactors and distillation towers	2.2
	Tracing flow patterns in oil/gas reservoirs	1.7
	Down-hole oil well studies, including corrosion inhibitor tracing, perforation marking, sand-fracture tracing	2.5

[a] Information provided by Dr J.S. Charlton (1999).

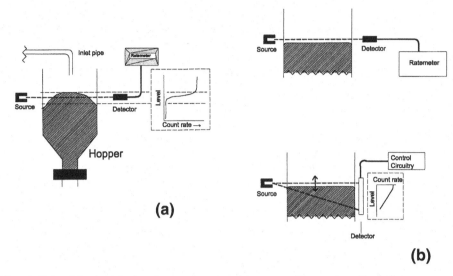

Figure 7.1. Nucleonic level gauge comprising a collimated radioactive source, a detector and a control system. (a) Monitoring the level of material in a hopper. The response of a fixed detector to the raising and lowering of the surface of the material is shown. (b) Monitoring the level of liquid in a tank. The gamma ray beam in the lower figure has an angle of about 20° and a linear detector is used (after Charlton, 1986, Ch. 13). The response of the detector to changes in the liquid level is shown.

7.2 Scientific and industrial applications of gamma rays

7.2.1 *Applications employing gamma ray attenuation*

Nucleonic gauges

Instruments used for the detection and measurement of nuclear radiation are often known as gauges and their use is referred to as gauging. A simple gauge comprises a radioactive source and a detector, e.g. a NaI(Tl) crystal with its signal processing equipment (Figure 4.2). An example is the level gauge shown in Figure 7.1.

A wide range of industrial applications of nuclear radiations is based on selected characteristics of γ rays. Several of these applications are summarised in Table 7.1 and include nucleonic level and density gauging, radiography, tomography and diagnostic column scanning. Relevant properties of γ radiations were discussed in Section 3.4.1.

Level gauges

Level gauges employ gamma ray attenuation in a particularly simple way and on a very large scale. They are widely used for monitoring or controlling the level of material in tanks or hoppers in the refining and chemical processing industries with many thousands installed around the world. Large plants employ numbers of gauges with information fed on-line to central operation rooms.

In its simplest form the gauge comprises a source of radiation in a specially constructed shielded housing and a detector. As shown in Figure 7.1(a), the response of the detector falls rapidly as soon as the surface level of a material in the hopper rises to intersect the beam. Factors affecting the choice of the radioactive source are discussed below.

In industry, there is often a need to record the position of the surface of a liquid in a tank over an extended range. To effect this, a more sophisticated version of the level gauge has been developed (Figure 7.1(b)) in which the radiation beam is collimated in the vertical plane with a dispersion angle of about 20°. A linear ionisation detector is used, the output of which is a measure of its length exposed to the radiations and therefore of the height of the liquid. To extend the height interval that can be monitored, multiple sources may be installed.

By feeding the detector output into circuitry connected to pumping and valving systems, the liquid levels may be automatically controlled within a pre-set range. This is the first of many examples of nucleonic control systems which are widely used in the control and optimisation of industrial processes.

Optimum choice of the radioactive source

The correct selection of the radioactive source involves matching the properties of available radionuclides to the specific application. For level gauges, the relevant properties are the energy of the emitted γ rays, the half life and the activity of the source.

(1) *The energy of the emitted γ rays:* The energy is chosen to ensure that the γ rays will readily penetrate the walls of the vessel, but that the beam intensity will be sufficiently attenuated by the material contained therein to permit accurate monitoring of the level.

   More exacting criteria govern the choice of sources for density gauges, γ radiography and tomography as these applications usually depend on monitoring small differences in the density of material with maximum precision. It can be shown that the maximum sensitivity is obtained when the product of the

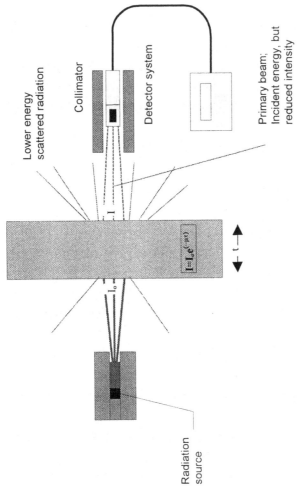

Lower energy
scattered radiation

Collimator

Detector system

Primary beam;
Incident energy, but
reduced intensity

Radiation
source

$I = I_0 e^{(-\mu t)}$

I_0

I

t

Figure 7.2.   The attenuation of a collimated gamma ray beam.

Table 7.3. *Half thicknesses of aluminium, iron and lead at selected γ ray energies.*

E_γ (keV)	Al[a]	Fe[a]	Pb[a]
60	9	0.7	–
100	15	2.4	–
300	26	8.1	1.5
600	33	12	5
1200	47	16	10

[a] The densities of aluminium, iron and lead are 2.70, 7.87 and 11.35 g/cm³ respectively. The μ_m values used to calculate these results were derived from Hubbell (1982). The $t_{50\%}$ results are uncertain by between ±5% and ±10%.

linear attenuation coefficient μ_t (cm⁻¹) and the thickness t (cm) of the sample along the beam path is unity, i.e. $\mu_t \times t = 1$ (Figure 7.2). In the Compton range, the linear attenuation coefficient decreases with increasing γ ray energy. Hence, the larger the sample, or the higher the density, the higher the gamma energy that is required for maximum sensitivity. In practice, for a uniform material, the energy of the gamma ray is chosen so that the path length of the collimated beam is close to the half thickness (or half value layer) (Table 7.3).

(2) *The activity of the source:* The source strength should not be greater than necessary so that the shielding requirements can be kept to a minimum. On the other hand, the activity of the source cannot be too weak. Otherwise the signal could be too low for a sufficiently rapid detector response. A correct choice for a level gauge is particularly important if the material being monitored can rise or fall quickly. Most practical applications can be designed with relatively weak and carefully shielded sources, so minimising radiation exposure.

(3) *The half life:* For installed nucleonic gauges the half life of a radionuclide used for an installation should be a year or preferably longer, since the longer the half life, the less frequently the source will need to be replaced. ⁶⁰Co, ¹³⁷Cs and ²⁴¹Am sources which are commonly used in level and density gauges have long half lives and well defined γ ray energies which together cover a wide range (see Table 8.1).

Density gauges

These gauges are used to measure the density of material between the source and the detector from measurements of the attenuation of a collimated γ ray beam (Figure 3.7(a)). The basic relationship between the measured γ ray intensity after passing the sample I_t, the initial intensity I_0 for a thickness t (cm) of the material of interest is given by

$$I_t = I_0 \exp\left(-\mu_m \times \rho \times t\right), \tag{7.1}$$

where the mass attenuation coefficient μ_m (cm^2/g) is related to the linear coefficient μ_t (cm^{-1}) and the density ρ (g/cm^3) of the material by $\mu_m = \mu_t / \rho$ (Section 3.4.4).

Density gauges differ from level gauges, which, in their simplest mode of operation, are just 'on/off' devices. Density gauges can detect differences in the density of materials as low as 0.1%. To achieve this level of precision it is necessary to employ a carefully shielded and collimated γ ray beam, and to optimise the selection of the source and, in particular, the energy and intensity of the emitted γ rays.

As indicated above, the γ ray energy of the source should be chosen to ensure that the beam is attenuated in the sample by approximately 50%. Under these conditions, small changes in density are measured with the greatest sensitivity, other factors such as counting time being equal.

To predict the thickness leading to a 50% reduction in the intensity of the beam, Eq. (3.8) (or Eq. (7.1)) is written $I_t = 0.5\ I_0$ and, on solving the equation, $\mu_t \times t_{50\%} \approx 0.693$,

$$t_{50\%} \approx 0.693 / \mu_t \approx 0.693/(\mu_m \times \rho_{av}), \tag{7.2}$$

where $t_{50\%}$ is the range of the γ rays in the sample leading to a 50% attenuation of their intensity.

Approximate values of μ_t and μ_m (in cm^{-1} and cm^2/g respectively) for the energies and materials of interest could be read from the graphs in Figure 3.7. Also, Table 7.3 lists $t_{50\%}$ for selected values of E_γ and a range of densities from that of aluminium to that of lead. When other values are required they can often be approximated by interpolation to sufficient accuracy. Alternatively, reference may be made to extensive tables and graphs of attenuation coefficients that have been posted on the Internet by the US National Institute for Science and Technology (NIST). Internet sources of nuclear data are discussed in Appendix 3.

An application of density gauges is to the monitoring of slurry densities in pipelines, shown schematically in Figure 7.3(a). Transmission gauges have also been used to monitor the deposition of scale on the walls of pipes (Figure 7.3(b)), since extensive scaling can significantly limit the capacity of the pipeline to transport material.

One of the major strengths of nucleonic gauges is that they may be used remotely and in hostile environments. Reference will be made to two devices developed by the commercial company ICI Synetix Tracerco for applications on offshore oil and gas platforms. Both are gamma transmission gauges which may be mounted by remotely operated vehicles (ROVs). The first, known as the *Gammagrout*^{TN}, system is designed to provide assurance that

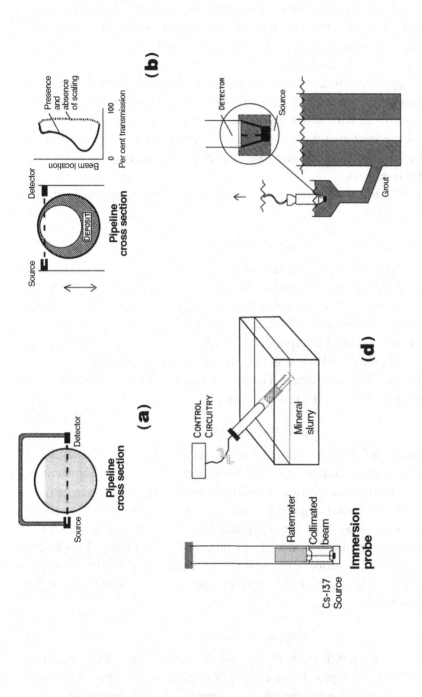

Figure 7.3. Four applications of gamma ray attenuation. (a) The monitoring of slurry in a pipeline. (b) Investigation of scale deposits in pipelines. (c) Monitoring of density of grout during the construction of, for example, offshore platforms (ICI Synetix Tracerco, product information). (d) The *in situ* measurement of the density of mineral slurries (Cutmore *et al.*, 1993).

the density of concrete grout used in the construction of piles supporting offshore oil platforms is within specification. Grout is pumped under pressure into the pile and continuously fills the space between the radioactive source and the detector. The density of the grout is directly related to the response of the detector which is monitored by personnel on the surface. Such a system is shown schematically in Figure 7.3(c).

The second application is known as the *Gammascan*TN *Flooded Member Detector System*. The principle is simple and robust. Flooding of a sub-sea member of a platform with water will reduce the transmission of the gamma beam between the source and the detector. The system is normally mounted on the arms of an ROV and is capable of rapidly monitoring both vertical and horizontal members.

Other applications of gamma ray transmission are discussed in the following paragraphs.

Mineral processing

Density gauges are widely used in the minerals processing industry. The real time analyses of the levels of valuable minerals in process streams generally employ radioisotope X ray fluorescence or γ ray preferential absorption techniques. As the responses of these detectors depend not only on the composition of the mineral component, but also on the bulk density of the aqueous slurry, nucleonic density gauges are an essential element of radio-isotope on-stream analysis systems. The gauge illustrated here (Figure 7.3(d)) is designed for immersion in the mineral slurry.

Coastal engineering

Nucleonic gauges have been adapted for the measurement of the levels of sediment in rivers and estuaries. Quantitative information on the mobilisation of sand and sediment under various conditions is important in the investigation phase of many coastal engineering projects. Either absorption or back-scatter gauges may be used for this purpose. They may be linked with other gauges measuring, for instance, depth, salinity and temperature. This information, together with position fixing data from satellite navigation systems, is integrated into computerised monitoring systems. Further comments are made in Section 9.3.5.

Radiography

Gamma radiography has long been one of the most important industrial applications of radioisotopes. The technology evolved out of the widespread use of X rays in medical imaging and has been adapted to monitoring the internal structure of manufactured components and to checking the integrity of welds.

The scientific principles of γ radiography are illustrated in Figure 7.4. Photographic X ray film housed in a suitable cassette holder serves to detect the transmitted radiation from the radioisotope source. If an inclusion in the sample under investigation has a higher electron density than the surrounding material, fewer γ rays are transmitted and a shadow image is produced on the film.

A wide range of radiography sources is commercially available. A selection is listed in Table 7.4. Factors determining the choice of the source are similar to those for density gauges as discussed above. Particular attention must be paid to the gamma ray energy. If the γ ray transmission is too high, the image contrast is inadequate. If the transmission is too low, there could be problems due to a low detector countrate. About 50% transmission is close to optimum. In practice, ^{60}Co, ^{137}Cs or ^{192}Ir sources are commonly used.

Computerised tomography (CT)

Computerised tomography (CT) is a sophisticated extension of radiography in which the detailed internal structure of a component can be obtained in two or three dimensions by an analysis of the attenuation of the X or γ ray beam in a large number of projections. Cormack and Hounsfield independently demonstrated the technique during the early 1960s (Bull, 1981) and were jointly awarded the Nobel Prize for Medicine and Physiology in 1979. However, it was not until the coming of sufficiently powerful computers during the early 1980s that tomography could be used routinely, first in medical imaging (Romans, 1995) and later in industry (Section 3.7.2).

When designing a CT instrument, a number of parameters need to be optimised. They include the energy of the incident beam, the data acquisition time and the resolution of the image. The criterion for energy selection is identical to that for density gauges. Data acquisition times should generally be as short as practicable to permit maximum utilisation of expensive equipment. Image resolution should be high, but there are intrinsic limits depending on the wavelengths (energies) of the radiation. Comparison of computed tomography for medical diagnosis and industrial inspection is discussed by Martz *et al.* (1990).

As with nucleonic density gauges and radiography, the principles of tomography are ultimately based on Eq. (3.7). This equation was derived to calculate the attenuation of γ rays in material made up of components differing in their linear attenuation coefficients μ_{ti} and thicknesses Δt_i. The sum $\sum \mu_{ti} \Delta t_i$ can be expressed as the line integral $\int \mu \, dl$. This is known as the

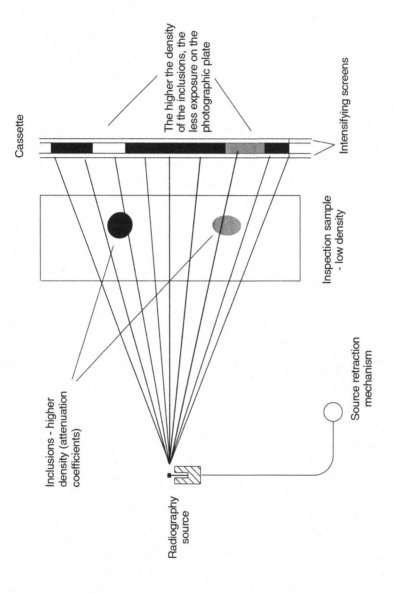

Cassette

The higher the density of the inclusions, the less exposure on the photographic plate

Intensifying screens

Inclusions - higher density (attenuation coefficients)

Inspection sample - low density

Radiography source

Source retraction mechanism

Figure 7.4. The principles of gamma radiography (after Charlton, 1986, Ch. 13).

Table 7.4. *Radioisotopes used as X or γ ray sources. (Based on Charlton, 1986, Table 14.1. For more details, see Table 3.2 of this book.)*

Isotope	Half life	Energy (keV)	Applications and comments
55 Fe	2.72 y	5.9 KX ray	Low energy X ray analysis. KX rays usefully excited Al to Cr
238Pu	87.8 y	12–17 LX ray	XRF applications. KX rays excited; Mn to Y
109Cd	463 d	22 KX ray	Detector calibration and XRF. KX rays usefully excited Fe to Mo
210Pb	22.2 y	47 KX ray	Used as an energy standard and for XRF
241Am	432 y	59.5 γ ray	Used for continuous gauging applications e.g. ash in coal measurements (Section 7.2.1);
			Used with target elements (e.g. copper or silver) to generate sources of (Section 7.2.3) pure fluorescent X rays XRF applications. KX rays usefully excited I to Lu
170Tm	127 d	84 KX ray	Low-energy radiography
57Co	272 d	122 and 136 γ rays	Mössbauer spectroscopy
T(^{3}H)/Zirconium	12.3 y	5 to 9	Bremsstrahlung sources. The X rays are generated as a consequence
147Pm/Aluminium	2.6 y	12 to 45	of the deceleration of the negatrons (β particles) emitted by the
90Sr/Aluminium	28 y	60 to 150	source (Section 3.8.1)

Radon integral after the mathematician who, in 1917, laid the foundation for later work on the mathematical reconstruction of tomographic images.

Tomographic systems have gone through several stages or generations of development. The principle is here illustrated using a first generation system shown in Figure 7.5 (Zatz *et al.*, 1981). Repeated measurements of the attenuation of a highly collimated beam of γ or X rays are made as the source–detector system traverses the sample. Similar data are then obtained at a number of projections by rotating the sample over a range of at least 180°. This information is accumulated by the computer which uses it to reconstruct a two or three dimensional image of the internal features of the sample.

These first generation systems produced excellent images free of artefacts. However, the systems were slow with data acquisition times sometimes upwards of several hours. More advanced and faster systems were designed to reduce imaging time by using an incident fan beam arrangement of X or γ rays and multiple detectors. Current instruments are based on those designed for hospital use. In a fully developed industrial system, more than a million measurements over 180° are used to develop a single CT image. The measurements can be made and the image reconstructed and displayed on a computer screen in about a minute. CT imaging has been applied to the study of machine components and to the condition of castings. It has been demonstrated that the CT-based measurements of the dimensions of complex castings are as accurate as those obtained by conventional means. However, for very complex or accurate work, imaging times ranging from a few minutes to a few hours may still be needed.

The size of the component that can be inspected depends on the energy of the incident beam. At one extreme, X ray sources from linear accelerators up to 15 MV in energy are being used commercially for a range of applications, including the examination of rocket motors.

At the other extreme, high-resolution micro-tomographic systems can provide information complementary to microscopy. Detailed information can be obtained on the internal structure of materials (such as wood and polymers) which would be destroyed during normal sample preparation. Using a low-energy X ray source, images with resolution of 5 μm have been obtained using small samples of wood (2 mm across). With this resolution, the cellular structure of the sample can be readily seen (Wells *et al.*, 1992).

High-speed systems are being developed in the steel industry for the real time gauging and control of products that are manufactured in a continuous process. This requires very fast computing capability able to generate output each 1 to 10 ms. Such systems have been demonstrated in the production of pipes and a range of structural materials with complex cross sections.

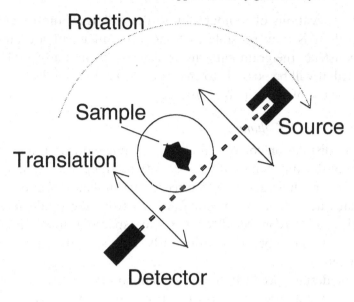

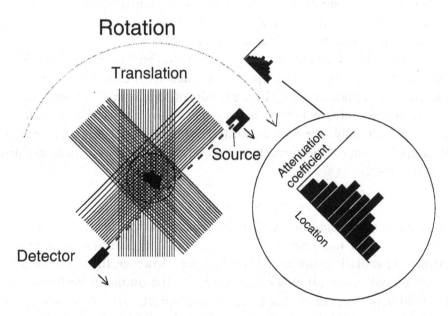

Figure 7.5. The scanning, rotation configuration for the first generation CT scanner (modified from Zatz, 1981).

Finally, on-line applications of tomography to the milling of timber are also being developed. It is now possible to tomograph an uncut log and mathematically reconstruct the grain patterns on veneers sawn at any angle. When fully developed, it will be possible to ensure that each log is milled in a way that optimises the use of the valuable resource.

Column scanning

The scanning of large distillation columns in the oil refining industry is a long established example of the application of γ ray density profiling. The columns are typically tens of metres high and two to three metres in diameter and are designed to separate the lighter petroleum fractions from the crude feed stock. They comprise an interconnected series of trays spaced at intervals of about one metre. Columns operate continuously and should require a minimum of maintenance.

Failure to meet the design specification of a column arises from a number of causes, notably the collapse of individual trays, the development of blockages leading to the flooding of trays and the formation of foams in the vapour phases. A number of these features are illustrated schematically in Figure 7.6. Many of these problems can be diagnosed using γ ray transmission techniques with no interference to plant operation.

The radioactive source and the detector are lowered in parallel on opposite sides of the column to develop a scan such as illustrated in the figure. However, a full understanding of the results of such scans in terms of column malfunction requires extensive experience. Many thousands of scans are performed annually, the vast majority by international service companies. Other applications of radiotracers in the oil refining industry will be discussed in Section 8.4.2.

On-line measurement of ash in coal

(1) *Dual isotopes applications:* The applications described in the previous sections involved monitoring the density and the distribution of material between the radiation source and the detector. However, there is often a need for additional information. For instance, for the optimum performance of coal fired boilers in the electricity generating industry, it is desirable that the feedstock be blended to a constant calorific value. Although the total ash and moisture contents must be known, the detailed elemental composition of the ash may be less important.

A widely used gauge employs two γ ray emitting radionuclides, ^{137}Cs (662 keV) and ^{241}Am (59.5 keV) as shown schematically in Figure 7.7. Attenuation of the 662 keV γ rays from ^{137}Cs is due mainly to Compton scatter. As shown

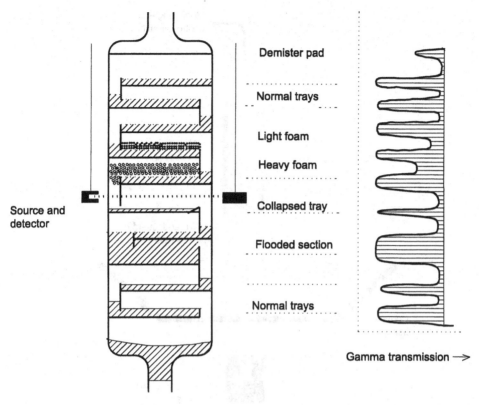

Figure 7.6. Gamma transmission scanning of a distillation column illustrating identifiable features (after Charlton, 1986, Ch. 13).

earlier (Section 3.4.6), Compton attenuation at that energy is almost independent of the composition of the material, so that the 662 keV γ rays can be used to monitor the mass of coal on the conveyor belt as it passes through the radiation beam.

By contrast, the attenuation of the 59.5 keV γ rays from ^{241}Am is due far more to photoelectric than to Compton interactions (Figure 3.7(a)) and hence increases rapidly with increasing concentrations of elements of higher atomic number such as iron and silicon (Eq. (3.3)). Coal heaped loosely on to conveyor belts is of fairly low density, permitting the transmission of some of the 59.5 keV γ rays through relatively thick layers. However, it may require ^{241}Am activities of order 10 to 20 GBq to ensure that the detected signal is strong enough for sufficiently precise measurements.

The percentage of ash in coal is calculated from a computer analysis of the transmission data of both the ^{137}Cs and the ^{241}Am γ rays. The overall uncertainty in the determination is normally in the range of 0.7 to 1.5 weight per cent (Cutmore *et al.*, 1993). Because the information is obtained in real

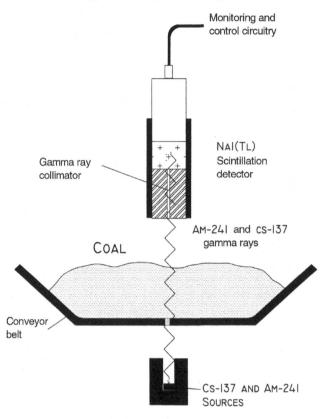

Figure 7.7. The 'on-belt' analysis of ash in coal using the dual-energy gamma transmission technique (after Cutmore *et al.*, 1993).

time, the outputs from a series of gauges installed on different conveyor belts can be used to optimise the blend of coal fed into the power station furnaces.

(2)   *Moisture in coke and coal, hybrid technology:* The on-line monitoring of the calorific value of coke and coal requires knowledge of the moisture content. Gamma transmission or backscatter gauges cannot be used to distinguish between the hydrogen in moisture and that associated with the organic component of the coal. One approach to overcome this difficulty has been to develop a gauge based on a combination of microwave phase shift and γ ray transmission techniques (Cutmore *et al.*, 1991).

(3)   *Multi element analysis of coal:* A still higher level of sophistication makes use of an elemental on-line analysis of the coal. Gauges have been developed commercially and are available from nuclear equipment manufacturers (Section 4.3.3) which can determine ash, moisture, density as well as the levels of key elements such as sulphur. The operating principles combine

prompt gamma ray neutron activation analysis (Section 7.4.4) with the dual transmission technology for ash and microwave methods for moisture.

7.2.2 Applications based on gamma ray backscatter

Backscatter gauges

In this section attention will be paid to the utilisation of γ ray backscatter. A number of applications are listed in Table 7.1.

Backscatter gauges are frequently used to monitor the levels of liquids in tanks when transmission measurements are not practical (Figure 7.8(a)). The intensity of backscattered radiation registered by the gauge depends primarily on the bulk density of the material in the tank. It is clearly greater below the liquid–gaseous interface than above.

The variation of the efficiency of single Compton backscatter (Figure 7.8(b)) with the angle of the scattered γ ray and with the density of the material in the tank is shown in Figure 7.8(c). The gauges are designed to optimise the backscatter angle. The countrate response increases with increasing density of the material and is enhanced by low-energy multiply scattered radiation.

The correct detection of the boundary between liquids of different densities depends also on the thickness of the tank wall. The walls attenuate both the primary beam and the backscattered radiation so reducing the sensitivity of the measurement. Backscatter measurements using γ rays at normally available energies are limited to vessels with wall thicknesses less than about 30 mm steel equivalent or thinner still when using low γ ray energies.

Backscatter techniques employing γ rays are not particularly suitable for locating liquid–liquid interfaces where the bulk densities of the two liquids are similar. This is different for neutron backscatter gauges (Section 7.4.2) which respond principally to differences in hydrogen concentrations. They are usually much more sensitive than γ ray detectors in defining the interface between immiscible liquids (Figure 7.8(d)).

Borehole logging using backscattered γ rays

A range of nuclear techniques, including γ ray backscatter, complement conventional methods in the comprehensive logging of boreholes (IAEA 1971, 1993). The essentials of a backscatter or γ–γ gauge, were introduced above and are illustrated in Figure 7.9(a). As discussed above, the gauge comprises a source and a detector separated by shielding to absorb directly transmitted radiation.

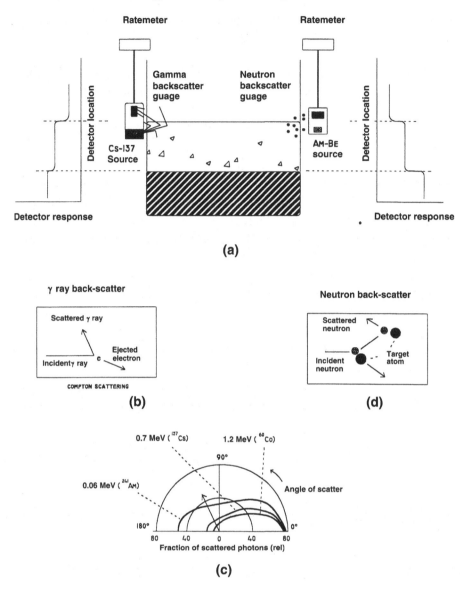

Figure 7.8: Gamma and neutron backscatter gauges. (a) Application of γ and n backscatter gauges to measurements of liquid levels in tanks. (b) Compton backscatter. (c) Variation with angle of the efficiency of single Compton backscatter for ^{241}Am, ^{137}Cs and ^{60}Co gamma rays. (d) Neutron scatter by elastic collisions.

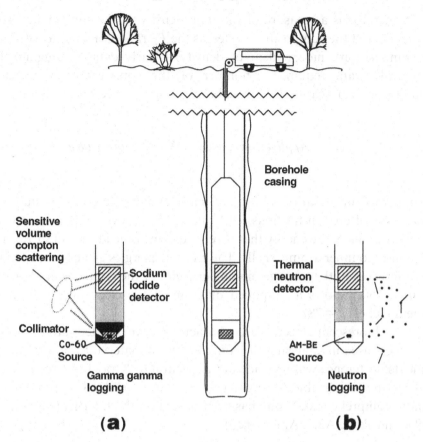

Figure 7.9.   Borehole logging: (a) γ–γ and (b) neutron backscatter logging techniques.

The response of γ–γ gauges depends primarily on the energy of the incident radiation. At energies above the range 100 to 300 keV (depending on the atomic number, Figure 3.7(a)), the Compton mechanism dominates and the detector response depends on the bulk density of the surrounds. At lower energies, the photoelectric effect becomes increasingly important and the more so the higher the atomic number of the surrounding material. Hence, the intensity of the low-energy component of the backscattered γ radiation is dependent on the effective atomic number Z_{eff}, i.e. on the composition of the surrounds.

In practice, measurements are made of the P_Z ratio which is given by given by:

$$P_z = \frac{\text{Intensity of Compton scattered radiation } (E_\gamma > 300\,\text{keV})}{\text{Intensity of radiation in the low-energy region of the spectrum.}} \qquad (7.3)$$

The numerator is a measure of the bulk density of the surrounding strata, independent of the composition, whereas the P_Z ratio is a measure of Z_{eff} (i.e. the composition), independent of density. Clearly detailed logging of the geological strata requires a number of the complementary techniques including neutron scatter which will be introduced in Section 7.4.2.

7.2.3 Applications based on X ray fluorescence

Introduction

Introductory information about the nature and properties of fluorescent X rays was offered in Sections 3.9.1 and 3.9.2. X rays emitted from excited atoms carry away the energy that is released when the atomic electrons move from outer to inner atomic shells. The emitted energies can be predicted from the differences in the average binding energies of the electrons in their shells. Figure 3.16(a) shows a simplified diagram of the electronic shells for the element nickel ($Z = 28$).

X ray fluorescent spectra are characteristic of excited atoms and are extensively used in analysing the composition of material for a wide range of elements. Reference will be made in this section to X ray fluorescence (XRF) analysis and to portable XRF gauges. X ray energies are listed in Table 3.2. A more comprehensive listing has been posted on the Internet (see Chu *et al.* (1999) and Table A3.1, Appendix 3).

X ray fluorescence analysis

X ray fluorescence analysis (XRF) is a technique for the analysis of a wide range of elements without the requirement for chemical pre-treatment. The sample is irradiated with an X ray beam of the required energy and the characteristic fluorescent X rays are detected (Figures 7.10(a) and (b)).

An X ray generator with a monochromator is the usual source since intensities are several orders of magnitude greater than the alternative radio-isotope sources. Monochromatic X rays yield increased analytical sensitivity by minimising the degree of overlap of the fluorescent X ray energies of interest with unwanted peaks from scattered X rays. The efficiency of emission of fluorescent radiation from an element reaches a maximum when the excitation energy is close to, but just above the absorption edge of interest.

There are basically two methods of detection: energy-dispersive X ray spectroscopy (EDS) using high-resolution semiconductor detectors or wavelength-dispersive X ray spectroscopy (WDS). Details are beyond the scope of

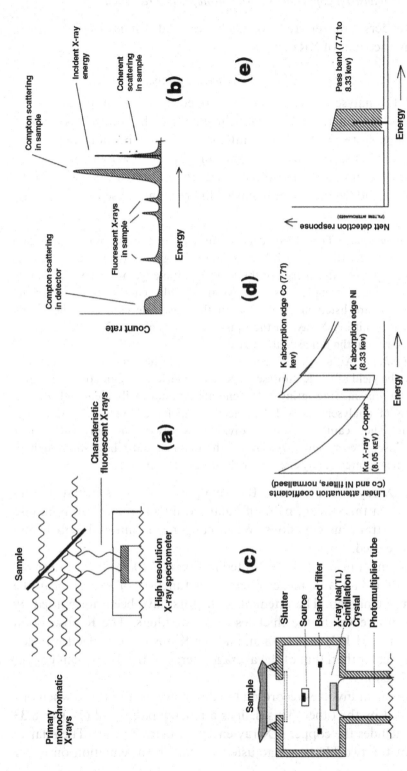

Figure 7.10. X ray fluorescence analysis. (a) Schematic representation of XRF analysis. (b) Features of an XRF spectrum (after Jaklevic *et al.*, 1977). (c) Schematic diagram of the head of a portable XRF analyser showing the location of the source, the sample, the balanced filter and the scintillation crystal. (d) The linear attenuation coefficients of the balanced cobalt and nickel filters in the vicinity of the copper K_α X ray emission. (e) The pass band, i.e. the change in the detector response when the balanced cobalt and the nickel filters are interchanged.

this text. Readers are referred to Lachance and Chaisse (1995) for a comprehensive account of XRF.

Portable X ray fluorescence gauges

Portable XRF analysers have been developed for field applications by replacing the X ray generator and monochromator with radioisotope sources and X ray filter systems. The lower intensity is offset by portability. The principle is here illustrated by the assay of copper. Two factors will be considered: (a) the choice of an isotopic source to optimise the emission of the copper KX rays and (b) the use of balanced filters to minimise interference by unwanted peaks.

(1) *Excitation energies:* The X ray energy emitted by the isotopic source and intended to trigger the copper KX rays should be greater than, but as close as practical to 9.0 keV, the energy of the K absorption edge of copper. A number of radionuclides emitting X rays or γ rays at a suitable energy are commercially available and are listed in Table 7.4. In this case one can use either ^{238}Pu ($T_{1/2} = 87$ y), emitting X rays in the range 12 to 17 keV or ^{109}Cd ($T_{1/2} = 1.2$ y) emitting X rays in the range 22 to 25 keV.

(2) *Balanced filters:* Although the correct choice of the excitation source will maximise the yield of the K fluorescent X rays, interference from other elements will also be present. Interfering X ray energies are not easily identified because portable XRF analysers use NaI(Tl) detectors which are of relatively low energy resolution. They cannot resolve individual KX rays but only show their average. Also, the entrance window of the detector must be thin enough to transmit the required X ray energies with sufficient intensity.

A portable X ray fluorescence (XRF) analyser is shown schematically in Figure 7.10(c). In this design, pairs of balanced filters are used to minimise the effects of extraneous emissions. When copper is monitored, cobalt and nickel filters are used.

The measurement principle is illustrated in Figure 7.10(d) which shows the attenuation for cobalt and nickel filters as a function of energy near the copper KX ray emission. The attenuation factors have been normalised by allowing for differences in the thicknesses of the filters. The K absorption edge for cobalt (7.71 keV) is less than, and the K absorption edge for nickel (8.33 keV) is greater than that of the average energy of the fluorescent copper KX ray (8.05 keV).

The measurement involves recording the difference in countrates when one filter is replaced by the other. This defines a narrow pass band (7.71 to 8.33 keV) which includes the copper KX ray energy (Figure 7.10(e)). The relative thicknesses of the two filters are adjusted so that the attenuation on either side of the pass band is balanced. By using the balanced filters, a good measurement of the level of copper is obtained in spite of potential interference from other elements in the sample which would not be resolved by the sodium iodide detector. A list of balanced filters of elements frequently used for KX ray detection is reported by Charlton (1986, Table 14.2).

Applications to the mineral processing industry

Reference was made above to the role of nucleonic density gauges in the real time analyses of the levels of valuable minerals in process streams. The identification of these levels is often made using radioisotope X ray fluorescence techniques. Since the complex minerals processing stream is the target, a complex spectrum of fluorescent X rays is produced.

In many industrial environments, the challenge is to monitor the desired economically important elements using ruggedised systems based on sodium iodide detectors which are of limited resolution. An example of such an approach is the use of balanced filters described above. A range of more elaborate techniques has been developed for applications in the minerals industry (see Watt, 1972, 1973).

7.3 Scientific and industrial applications of beta particles and electrons

7.3.1 Attenuation of beams of beta particles and electrons

Whereas γ rays are uncharged electromagnetic radiations, β radiations are charged particles interacting with matter primarily through Coulomb interaction with the outer electrons of its atoms (Section 3.3.1). The range, t cm, of a beam of mono-energetic electrons depends principally on the energy of the electrons and the density, ρ g/cm^3, of the material in which they move. It was shown in Section 3.3.5 that the product $t \times \rho$ g/cm^2 known as surface density, has the useful property of being only weakly dependent on the Z number of the absorbing element. With β particles having short ranges in solids (Figure 3.2(b)), their attenuation is used for the measurement of the thickness of films in the paper, plastics and rubber industries. Comprehensive listings of the ranges of electrons have been published in regular reports (Berger and Seltzer, 1964) and on the Internet (Berger, 1999).

Frequently used commercially available β particle sources include ^{85}Kr ($T_{1/2} = 10.73$ y, E_β(max) 672 keV) and ^{147}Pm ($T_{1/2}$ 2.62 y, E_β(max) 225 keV) (Table 8.2). The detectors are saturation ionisation chambers filled with argon at two or more atmospheres pressure to increase their efficiency (Section 6.3.4). Ionisation chambers are used because they can cope with high particle fluxes without the need for dead time corrections (Section 4.5.2).

Applications in paper manufacture

Nucleonic gauges using such sources are fitted routinely to new paper manufacturing plants and are often retrofitted to older units. They may be used to continuously monitor the thickness of paper at speeds up to 400 m/s. Optimum ranges depend on the source chosen (0.03 to 1 g/cm^2 with ^{85}Kr, and 0.015 to 0.175 g/cm^2 with ^{147}Pm).

On-line processing of the output data from the detectors allows the collection and display of short-term trends and overall average variations in the thickness of the paper. The output data may be used to control the speed and inclination of the rollers to ensure that the paper thickness always remains within tolerance. Using microwave or infrared techniques, the moisture level is monitored to about 0.2% and controlled by automatically varying the amount of evaporative heating. Through the use of these control systems, the paper is manufactured within tighter tolerances than would otherwise be possible so gaining a higher quality product and a reduction in the use of materials and energy.

7.3.2 Industrial applications of beta particle backscatter

Assuming saturation thickness (Section 3.3.3), the efficiency of the back-scatter of β particles increases with the energy of the incident radiation and with the electron density i.e. the effective Z number of the backscattering material (Figure 3.5).

While β particle transmission techniques are extensively used in industry to measure the thickness of paper and other films, β particle backscatter methods are well suited to measurements of the thickness of thin coatings on substrates that are thick enough for saturation backscatter (Section 3.3.3). The radiation source in a backscatter gauge is a pure β emitting isotope with an energy chosen to suit the application. Examples include ^{63}Ni, ^{147}Pm and ^{204}Tl (Table 8.2). A practical gauge incorporates a micro-processor to convert the reading into a meaningful measure of coating thickness.

Since the efficiency of backscatter from a material is proportional to its electron density, the best defined applications are those in which a material of high atomic number is laid on one of preferably much lower atomic number or vice versa (Charlton, 1986, Ch. 14). Applications include the monitoring of the thickness of electroplated gold and the thickness of plastic coatings on metals (see Table 7.5); suitable sources for these examples are ^{204}Tl and ^{63}Ni respectively.

7.3.3   *Special applications: electron microscopy*

Knowledge of the relationships between the properties, structure and compatibility of materials under a wide range of conditions underpins many of the processing, manufacturing and heavy engineering sectors of industry. Electron microscopy, which involves the application of many of the scientific principles discussed in this book, has been used to advance knowledge in these areas. In simple terms, an electron microscope comprises a source of electrons and a series of electro-magnets, which perform functions similar to that of lenses in optical microscopes.

The fundamental difference between optical and electron microscopes is the mechanism of formation of the contrasts resulting in the image. In light optics, images result from differences in absorption of the light illuminating the object; in electron microscopy they arise from the complex processes of electron scattering and diffraction.

In both types of instruments, the Rayleigh criterion requires that the limit of resolution, d, is a function of the incident wavelength given by $d = k\lambda/a$ where k is a constant, λ is the wavelength of the incident radiations and a is the aperture of the microscope, all in consistent units. For optical microscopes, the resolution lies in the range 0.2 and 0.5 µm (2000 to 5000 Å), leading to a maximum magnification of around 1000 ×.

The wavelengths of electrons decrease as their energies increase i.e. with increasing accelerating voltage. Electron wavelengths are very short, only 0.09 Å at 20 keV and 0.025 Å at 200 keV, and so substantially less than the average interatomic spacings which are about 2 Å. With modern high-resolution techniques, well maintained and properly aligned transmission electron microscopes can resolve at the atomic scale. However, due to a number of instrumental effects, the resolution for scanning electron microscopes is limited to between 50 and 200 Å, which corresponds to magnifications of about 100 000 ×.

The interaction of higher energy electrons (> 5 keV) with the atoms of the specimen leads to the generation of X rays. These comprise both bremsstrahlung (Section 3.8.1) and fluorescent X rays (Section 3.9.1). The latter are used to map the distribution of elements in the sample. In addition, structural information may be obtained through the analysis of electron diffraction patterns. Readers seeking further information are referred to works by Hunter *et al.* (1993), Williams and Carter (1996) and Watt (1997).

Table 7.5. *Industrial applications of beta particles and electrons.*

Property of the Radiation	Application	Example	Reference
Transmission	Thickness measurements	Thickness control in paper manufacturing industry based on the attenuation of transmitted ^{85}Kr β particles	Section 7.3.1
Backscatter	Thickness measurements of coatings		Section 7.3.2
Absorption	• Chemical processing; • Sterilisation; • Radiation curing.	Irradiation by electron beams or the derived bremsstrahlung is an alternative to gamma ray treatment. Electron beam irradiation complements UV as a curing agent in the printing and coating industries	Section 7.6
Special applications	Electron microscopy	Investigation of materials characteristics, composition and structure	Section 7.3.3

Note: The properties of beta particles are discussed in Section 3.3 and the decay data for sealed beta particle sources are presented in Table 8.2.

7.4 Scientific and industrial applications of neutrons

7.4.1 Comments on work with neutrons and neutron doses

Neutrons are uncharged sub-atomic particles with a mass similar to that of protons (Table 1.2). They are of industrial and scientific importance because their applications contribute knowledge of the structure and composition of materials not readily available from other techniques (L'Annunziata, 1998, Ch. 1).

Properties of neutrons and of selected neutron sources were briefly described in Sections 1.3.6 and 5.4.4. Applications employing neutrons will be introduced with a few comments on estimating dose rates to experimenters working with neutrons. For full information see, for example, Cember (1996, Chs. 5 and 6).

Table 2.1(a) lists non-dimensional radiation weighting factors w_R (see Table 2.2 for definitions), required to calculate equivalent doses due to nuclear radiations absorbed by operators (Eq. (2.4a)). Table 2.1(a) shows that weighting factors for neutrons, $(w_R)_n$, are functions of the neutron energy. They reach a maximum of 20 for neutron energies between 0.1 and 2 MeV, when they are equal to w_R for α particles. However, they decrease to 5 at higher energies and also for thermal neutrons. Nevertheless, the fact that $(w_R)_n$ is at least five times larger than $(w_R)_\gamma$, the weighting factor for γ rays, reflects the potentially strong interaction of neutrons with tissue material.

Protection against high-flux neutrons in research reactors is provided by thick walls of metal and concrete (Section 1.4.1). Portable sources of the type described in Section 1.3.6 (Figure 1.6(a)) are designed to ensure that doses to the operators are well within the international guidelines and are as low as reasonably achievable (Section 2.7.2 in this book and Martin and Harbison, 1996, Section 8.5).

7.4.2 Industrial applications of neutron sources

Neutron sources

There are three sources of neutrons which are of importance for industrial applications: nuclear reactors, neutron generators and portable isotopic sources. The neutron flux densities are in the ranges 10^{12} to 10^{15} n/(cm^2 × s) for reactors, 10^6 to 5×10^{11} n/(cm^2 × s) for accelerators and 10^4 to 10^7 n/s for portable neutron sources.

The energies of moderated neutrons from reactors are predominantly in the thermal and epithermal range, while those from the other sources are in the megaelectronvolt range. Neutrons may be classified according to their energy. A scheme based on the response of commonly used detectors is included in Section 5.4.4. For a different classification of neutron energies see, e.g. L'Annunziata, (1998, Ch. 1). A list of applications of neutrons in science and industry is given in Table 7.6. The three sources mentioned at the start of this section are now discussed in turn.

(1) *Research reactors:* Reactors produce an intense flux of neutrons which are employed principally for:
   • the production of radioisotopes for medicine, industry, agriculture and scientific research;
   • the neutron activation analysis of materials recovered during geological research, minerals exploration and environmental monitoring;
   • the neutron transmutation doping of silicon for the semiconductor industry;
   • the structural determination of materials using neutron diffraction;
   • neutron radiography.

(2) *High-voltage neutron generators:* Deuterium–tritium sealed tube generators produce pulses of 14 MeV neutrons and are widely used in sophisticated borehole logging applications. Ionised deuterium gas molecules are accelerated to 160 kV to bombard a mixed deuterium/tritium (say 560 GBq or 15 Ci) target and generate neutrons via the $^3H(d,n)^4He$ reaction. A typical down hole accelerator produces a flux of about 3×10^8 n/(cm^{2}s) with a thermal/fast flux ratio of about 1% and a pulse length of, say, 16 μs.

(3) *Portable neutron sources:* The portable sources in most common use employ neutrons that are emitted: (a) when α particles interact with the atoms of light elements (commonly alphas from americium-241 or plutonium-239 interacting with beryllium atoms and (b) following the spontaneous fission of californium-252. A third class of source exploits the interaction of gamma rays from antimony-124 with either heavy water (D_2O) or beryllium. The reactions are listed in Table 7.7. The reader is referred to IAEA (1993) for further details.

The energies of the emitted neutrons are mostly in the range 2 to 10 MeV (Figure 1.6(b)). The Am/Be, Pu/Be and ^{252}Cf sources are quasi-point sources, are compact and transportable and have been incorporated into a range of nucleonic instruments. These include:

• industrial backscatter gauges for the measurement of liquid surfaces and interfaces in tanks and reaction vessels;
• soil moisture meters for use in agriculture, hydrology and civil engineering;
• borehole logging tools.

Industrial applications are classified according to the property of the neutrons

Table 7.6. *Applications of neutrons to science and industry.*

Property of the Radiation	Application	Example	Reference
Backscatter	Measurements of differences in the moderating power (effectively the hydrogen content) of materials	(a) Monitoring liquid levels in tanks (b) Measuring moisture in soils, geological strata and bulk materials. There are specific applications to agriculture, civil engineering and borehole logging (see below)	Section 7.4.2
	Neutron–neutron borehole logging	(a) Porosity measurements (b) Monitoring the freshwater/saline contact zone in coastal aquifers and the water/oil interface zone in oil wells	
Diffraction	Materials structure	Neutron diffraction, being particularly sensitive to H and the light elements, complements X ray techniques	Section 7.4.3
Neutron activation–delayed	Neutron activation analysis; Production of radioisotopes	Multi element assay; Radioisotopes used in nuclear medicine, industry, agriculture	Section 7.4.4 Section 1.4.4
Neutron activation–prompt gamma	Borehole logging	Obtain data on the elemental composition of the surrounding strata, and hence information on the lithology and groundwater salinity	Section 7.4.2
	On-line analysis	Monitor the level of sulphur and the elemental composition of ash in coal on conveyor belts	Sections 7.2.1 and 7.4.4

Note: The properties of neutrons are discussed in Sections 1.3.6 and 5.4.4.

Table 7.7. *Radioisotope neutron sources (Shani, 1990).*

Neutron source	Neutron yield (n/s/Ci[a])	Type of source	Half life	Reaction
Pu–Be	1.7×10^6	α/n	24,360 y	$^9\text{Be} + \alpha \rightarrow {}^{12}\text{C} + \text{n} + 5.71$ MeV
Am–Be	2.2×10^6		433 y	
$^{124}\text{Sb–Be–D}_2\text{O}$		γ/n	60.2 d	$^9\text{Be} + \gamma \rightarrow 2\alpha + \text{n} - 1.67$ MeV
				$^2\text{H} + \gamma \rightarrow {}^1\text{H} + \text{n} - 2.23$ MeV
$^{252}_{92}\text{Cf}$	4.4×10^9	Spontaneous fission (SF)	2.65 y effective	$^{252}_{92}\text{Cf (SF)} \rightarrow 2\text{f} + 3.8\text{n} + 200$ MeV (Section 5.4.4)

[a] One curie (Ci) is 37×10^9 Bq i.e. 37 000 MBq.

that is exploited, namely scattering, activation or diffraction. A few comments on radiation protection were included in Section 7.4.1.

Neutron moderation

Neutrons from portable sources are generated at megaelectronvolt energies. On interacting with surrounding matter they dissipate their kinetic energy, principally by elastic and inelastic scattering processes. Ultimately they equilibrate thermally with their surroundings with energies of about 0.025 eV near room temperature. When configured as a backscatter gauge, the neutrons are detected by a calibrated proportional counter placed adjacent to the neutron source (Section 5.4.4). The intensity of the neutrons backscattered into the proportional counter is determined principally by the hydrogen concentration in the material of interest, though there are some complicating factors yet to be mentioned. The process by which kinetic energy is dissipated is known as neutron moderation.

Elastic scatter conserves the initial kinetic energy of the neutrons, but redistributes it as in billiard ball collisions. The energy transfer from the neutrons is greatest for collisions with hydrogen atoms, since protons and neutrons are of equal mass. The probability for energy transfers decreases with increasing mass of the target nuclei. In water, fast neutrons are thermalised in a distance of about 10 cm, predominantly by elastic scatter with hydrogen atoms. The energy of the recoiling protons is dissipated by ionisation and excitation of the atoms and molecules of the surrounding material. Having reached thermal energies, the neutrons are captured, as a rule by hydrogen atoms to form deuterium with the emission of 2.223 MeV γ rays. This is an example of a neutron capture followed promptly by gamma ray emission.

Inelastic scatter occurs when the neutrons lose some of their kinetic energy to form metastable states in nuclei. Normally, the excited nuclei promptly de-excite to the ground state with the emission of γ rays and possibly nucleons. The element in question can often be identified from the energy of the emitted γ rays. This is the basis of prompt gamma neutron activation analysis (PGNAA) which will be discussed in more detail in Section 7.4.4. Inelastic scattering will only occur when the energy of the neutron is above a threshold value. The values vary from element to element but are generally in the range 0.5 to 7 MeV. The emission of an inelastic scattering gamma ray is not necessarily a prompt process. An interesting example is the excitation of rhodium-103 to a metastable state rhodium-103m ($T_{1/2}$ 56.12 min), which de-excites with the emission of 40 keV γ rays.

The efficiency with which fast neutrons dissipate their kinetic energy (elastically and/or inelastically) and slow down is known as the moderating power of the surrounding medium.

Neutron backscatter gauges

A neutron backscatter gauge comprises a source of fast neutrons (commonly Am/Be) and a proportional counter using helium-3 or boron trifluoride which responds only to backscattered neutrons down to thermal energies (Figure 7.8(a), see also Charlton, 1986, p. 278). Thermal and epithermal neutrons can be detected by helium-3 proportional counters whereas boron trifluoride detectors are sensitive only to thermal neutrons. Both detectors record an almost zero response to fast neutrons. Hence, as noted earlier, unshielded proportional counters can be placed adjacent to the fast neutron source without recording a significant background reading (Section 5.4.4).

Neutron backscatter gauges are routinely used in the oil refining and chemical industries to monitor the levels of liquids in tanks and to detect the interfaces between different liquids. With hydrogen being a highly efficient neutron moderator, the rate of detection of backscattered, thermalised neutrons is used as a measure of the concentration of water or hydrocarbon in the tank. For the arrangement shown in Figure 7.8(a), it is possible to monitor not only the level of the surface of the oil, but also its interface with the denser water. As shown schematically in the figure, it would be more difficult to accurately locate such an interface with an externally located γ backscatter gauge which responds principally to differences in bulk density.

Neutron moisture meters

Soil moisture meters are specialised backscatter gauges. They comprise a source of fast neutrons and a detector responding only to thermalised neutrons configured into a portable field instrument. The probe is lowered into the borehole when the countrate can be related to the moderating power of the soil in the sphere of influence around the probe which is typically of the order of 0.5 m radius. The countrate is usually normalised to that observed when the probe is suspended in a water-filled portable standard drum.

The moderating power of the soil is determined principally by its total hydrogen content. Both soil moisture and organic hydrogen are present. A calibration step employing an oven dried sample of soil to monitor the level of bound hydrogen [H] is necessary before the water content can be deduced from the neutron meter readings. Additional corrections are needed for the percentage of thermal neutrons which are absorbed by other components of the soil (see below), and are therefore not recorded by the detector (Longworth, 1998, Ch. 5). Quantitative estimates are based on the concept of the macroscopic thermal neutron absorption cross section Σ_a, which is defined as

$$\Sigma_a = n_x\sigma_{a,x} + n_y\sigma_{a,y} + n_z\sigma_{a,z} + \cdots \tag{7.4}$$

where n_x, etc. is the atom fraction of element x in the soil and $\sigma_{a,x}$, etc. is the corresponding neutron absorption cross section. Values of n_x are obtained from the chemical analysis of the soil and $\sigma_{a,x}$ from tabulated values (see e.g. Figure 1.3) or via the Internet (e.g. http://t2.lanl.gov/data/map.html).

Problems may arise because very minute traces of strong neutron absorbers can have a significant effect on Σ_a and are not readily assayed with sufficient accuracy. One approach is to measure the response of the neutron moisture meter to soil of known values of the hydrogen concentration [H] and Σ_a in a standard calibration drum.

Moisture gauges are widely used in agricultural research for the monitoring of soil moisture variations in the root zone. They have also found widespread use

- in civil engineering to measure the moisture levels in bulk material employed for the construction of roads and earth-filled dams;
- in the concrete and glass industries to monitor the moisture levels in sand; and
- in the iron and steel industry to determine the moisture levels in coke and sinter mixtures.

Borehole logging with neutrons

The scientific principles underpinning the neutron moisture meter have been extended and applied to the logging of oil fields and groundwater bores (Figure 7.9(b)). The purpose of this section is to indicate the level of sophistication resulting from decades of development, principally in the oil industry. Major applications include measurements of the

hydrocarbon volumes and viscosity;
porosity and permeability; and
grain size and mineralogy.

Reference will be made to the measurement of hydrocarbon volumes and of porosity.

(1) *Hydrocarbon volumes:* Hydrocarbon volumes in oil bearing strata are normally obtained by conventional borehole logging. However, where there is a need for fast, accurate measurements of the hydrocarbons prompt γ neutron activation analysis (PGNAA) methods are used (Section 7.4.4). In practice it is the carbon/oxygen ratio which is monitored. This ratio varies from zero in the absence of hydrocarbons to a maximum value when the surrounding strata are saturated

with hydrocarbons. Complications arise when the petroleum occurs in carbonate rocks such as calcite [CaCO$_3$] and dolomite [CaMg(CO$_3$)$_2$].

(2) *Porosity:* Porosity is the fraction of the total volume occupied by pore space and is an important parameter in the evaluation of the oil bearing potential of a formation. The porosity can be determined from the average bulk density if the average matrix density is known. The bulk density can be estimated using γ ray backscatter (Section 7.2.2). The matrix density, which is the physical density of the material comprising the formation, can be calculated from knowledge of the mineral abundances.

In practice, there is a very limited range of minerals commonly found in many oil bearing strata. It is therefore often possible to infer the mineral composition from multi element analyses. Such analyses can be obtained, for instance using PGNAA techniques, supplemented by estimates of potassium, uranium and thorium levels obtained from γ ray logging.

Direct estimates of porosity can be obtained from measurements of the lifetimes of epithermal neutrons (Wilson *et al.*, 1989). A down hole deuterium–tritium (D–T) generator is used to produce pulses of 14 MeV neutrons with a pulse time of about 16 μs and a repetition rate of 5 kHz (i.e. every 200 μs). The response of the detector to epithermal neutrons is recorded at microsecond intervals. The energies of the neutrons are moderated by the surrounding strata. The greater the porosity of the strata, the greater the level of hydrocarbons and hence the greater the moderating power of the surroundings and the shorter the lifetime of the epithermal neutrons.

To apply this technique it is necessary to monitor the energy loss of epithermal neutrons free from thermal neutrons. This is achieved by fitting a ^{3}He detector with a gadolinium cover to absorb the thermal component.

Neutron radiography

This technique uses neutrons to obtain information about the internal structure and distribution of material under investigation. As with γ ray radiography (Section 7.2.1) the neutrons produce an image on a photographic film (via protons scattered by the neutrons out of hydrogen atoms), which reflects the pattern of opaqueness of the object to the radiation. However, the mechanisms of generating the images are different.

On the one hand, a γ radiograph records variations in the linear attenuation of γ rays within the object. On the other, the response of the photographic film to an incident neutron beam reflects differences in the efficiencies of scattering and absorption processes. The greater the extent of scattering or the greater the absorption of neutrons, the less the response on the film. Materials with a high hydrogen content or with a high neutron absorption cross section Σ_a will be seen as regions of high opaqueness on the film. There are many occasions when gamma and neutron radiography can provide complementary information.

7.4.3 Neutron diffraction

The French physicist Louis de Broglie postulated, back in 1924, that according to quantum theory, elementary particles can, in appropriate circumstances, be diffracted like waves. It can be shown that the wavelength λ of a stream of particles, each of mass m and velocity v, is inversely related to its momentum p $(=mv)$ with the Planck constant h as the constant of proportionality:

$$\lambda = h/p = h/mv, \tag{7.5}$$

all in consistent units. When applied to neutrons, Eq. (7.5) can be written

$$\lambda = h/(2m_n E_n)^{1/2}, \tag{7.6}$$

where E_n is the kinetic energy of the neutrons. For thermal neutrons one has $m_n = 1.67 \times 10^{-27}$ kg, $E_n = 0.025 \times 1.6 \times 10^{-19}$ J and, also, $h = 6.63 \times 10^{-34}$ Js. Substituting into Eq. (7.6) one obtains

$$\lambda = 1.8 \text{ Å}, \tag{7.7}$$

which is of the order of interatomic distances in numerous materials at room temperature (Hey and Walters, 1987, Ch. 3). Neutron wavelengths, within the range 0.1 to 20×10^{-10} m, are accessible. Longer wavelengths are produced by cooling the neutrons, e.g. to liquid helium temperatures (≈ 4K), so lowering their kinetic energies and increasing λ (Eq. (7.6)). Shorter wavelength neutrons are generated using hot graphite at up to 2400 K with a corresponding increase in kinetic energy. With these capabilities, it is possible to employ neutron diffraction for investigations of the structure of solids with dimensions ranging from atomic spacings to those characteristic of large polymers or biological molecules.

Neutron diffraction techniques complement conventional diffraction methods using X rays. X rays of similar wavelengths to neutrons interact principally with atomic electrons and therefore preferentially with high Z number atoms. In contrast, neutrons interact preferentially with low Z number nuclei. Neutron diffraction techniques are more useful than X rays in locating the position of hydrogen atoms in samples and are widely employed for the investigation of the structure of biological materials. The technique has also been used to study the crystal structures of new magnetic and semiconducting materials which are being developed for advanced communications.

7.4.4 Neutron activation analysis (NAA)

An overview

Neutron activation analysis techniques are extensively used in geology, geochemistry and environmental science for multi elemental analysis. NAA involves: (a) irradiation of the sample in a research reactor (Sections 1.3.6 and 1.4.1), or with a neutron generator or portable neutron source, (b) detection and measurement of the level of radioactive isotopes resulting from the neutron activation of the target material (Section 1.4.3); and (c) calculation of the concentration of each activated component of the sample.

The energies of the emitted γ rays have to be measured using a high-resolution γ ray spectrometer to distinguish between the often large number of activated nuclides. The sensitivity of the analysis depends on the neutron absorption cross section of the isotopes of interest, the neutron flux density and the effect of interference from other elements in the sample. Minimum detectable levels for some 90 elements when using NAA in optimum conditions are listed by Charlton (1986, Table 15.2).

Prompt neutron activation analysis

Neutron interactions with numerous elements lead to prompt (n,γ) reactions (Section 7.4.2). The analytical technique based on measuring the energies and intensities of the emitted γ rays is known as prompt gamma ray neutron activation analysis. Measurements must be made during neutron irradiation. They cannot therefore be undertaken within the core of a nuclear reactor and are restricted to portable neutron generators or isotopic sources (Table 7.7). However, they can be applied when the half lives of the daughters are either too short or too long for delayed or normal NAA.

PGNAA is widely used in borehole logging (Section 7.4.2) where it is necessary to monitor a wide range of elements including hydrogen, carbon and oxygen (IAEA, 1993). Quantitative and qualitative aspects of neutron–gamma borehole logging have been discussed by Charbucinski *et al.* and by Mikesell *et al.* in IAEA (1990b). Some gamma ray energies listed by the latter group are presented in Table 7.8 for illustration.

PGNAA is also routinely applied to monitoring the elemental composition of coal and other materials on conveyor belts (James, 1990). As noted in

Table 7.8. *Energies of major (γ s/n >5%) photopeaks from selected elements (after Mikesell et al. in IAEA, 1990b).*[a]

Element	Photopeak energy (MeV)
Silicon	3.539, 3.028 (se[b] of 3.539), 2.517 (de[c] of 3.539), 4.934, 4.423 (se of 4.934), 3.912 (de of 4.934) 2.093
Aluminium	1.779, 7.724, 7.213, 6.702,
Iron	7.631, 7.120 (se[b] of 7.631), 6.609 (de[c] of 7.631), 7.645, 7.134 (se of 7.645), 6.623 (de of 7.645), 5.920, 5.409 (se of 5.920), 4.898 (de of 5.920), 6.018 , 5.507 (se of 6.018), 4.996 (de of 6.018)
Calcium	1.942, 6.420, 5.909 (se[b] of 6.420), 5.398 (de[c] of 6.420), 4.419, 3.908 (se of 4.419), 3.397 (de of 4.419)
Nickel	8.999 , 8.488 (se[b] of 8.999), 7.977 (de[c] of 8.999), 8.533 , 8.022 (se of 8.533), 7.511 (de of 8.533), 0.465, 6.837, 6.326 (se of 6.837), 5.815 (de of 6.837)
Manganese	0.847, 1.811, 7.244, 6.733 (se[b] of 7.244), 6.222 (de[c] of 7.244), 7.058 , 6.547 (se of 7.058), 6.036 (de of 7.058)
Magnesium	3.917, 3.406 (se[b] of 3.917), 2.895 (de[c] of 3.917), 2.828, 1.809, 0.585
Sodium	0.472, 6.395, 5.884 (se[b] of 6.395), 5.373 (de[c] of 6.395), 0.871, 3.982, 3.471 (se of 3.982), 2.960 (de of 3.982), 2.027
Potassium	1.461 (natural decay), 0.770, 5.381, 4.870 (se[b] of 5.381), 4.359 (de[c] of 5.381)

[a] The table is designed to indicate the range of prompt gamma neutron activation peaks. In practice many of the peaks are not used because of interferences, or because they lie outside the spectral range of the instrumentation.
[b] se single escape peak (Section 6.4.2).
[c] de double escape peak.

Section 7.2.1, these data may be used to supplement on-line measurements of ash in coal.

Instrumented neutron activation analysis

Nowadays, instrumented neutron activation analysis (INAA) is used where possible to cope with the growing demand for measurements. INAA involves (a) the automatic transfer of the sample to the reactor irradiation position for a pre-determined time, (b) subsequent counting of the activity of the sample using a high-resolution solid state detector and (c) the automated analysis of the resulting γ ray spectrum in terms of the concentrations of the required elements. The method is non-destructive but there is a limit to the range of elements that can be measured. The shorter the half life of the element, the shorter has to be the cycle for sample transfer operations, activation and counting. The minimum half life which can be accommodated in favourable conditions is a few seconds.

Other comments

Activation analysis has been applied in numerous fields, including the analysis of biological materials, coal and coal effluents, water samples, air particulate matter and semiconductor materials. It has also been applied to *in vivo* studies, archaeology, agriculture and botany. Further information may be obtained from Alfassi (1990) or Kruger (1971).

The large majority of NAA measurements use radioactivation and so are carried out most efficiently by thermal neutrons. Epithermal or fast neutron irradiations (Section 5.4.4) are employed when reduced efficiency of activation is compensated by reduced interferences. Many nuclear reactions will only respond to fast neutrons. The intensities of epithermal or fast neutrons are measured by using neutron filters such as cadmium sheets to filter out the thermal flux. Reactions triggered by fast neutrons, e.g. (n,p) or (n,α) reactions, are also used to measure the energies and intensities of fast neutrons in the reactor spectrum.

For assays of highest sensitivity, neutron activation techniques are combined with either chemical concentration of the target element prior to irradiation or radiochemical separation following the activation of the sample (Fardy, 1990).

7.5 Scientific and industrial applications of protons and alpha particles

7.5.1 Introduction

High-energy protons or alpha particles impinging on a target material will form a range of reaction products including X rays, gamma rays and neutrons (Peisach, 1990). Their properties underpin a number of applications which may be broadly classified as

multi element analyses
thin layer activation, and
general applications such as smoke detection.

The implantation of metals and alloys with heavy ions can lead to surface hardening which is exploited in industry. However, a detailed discussion is beyond the scope of this text.

7.5.2 Multi element analyses

The irradiation of a target material by protons or alpha particles leads to the excitation of the atoms and their nuclei with the resulting emission of

X rays and gamma rays. The energies of the emitted radiations are indicative of the elements present and the intensities reflect their concentrations. These properties have been exploited over many years for the multi element analyses of natural materials. The most common arrangement involves the activation of the sample material with a proton beam from a Van de Graaff accelerator and the measurement of the emitted X rays and gamma rays. These techniques are known as proton induced X ray emission (PIXE) and proton induced gamma ray emission (PIGME) respectively. The beams may be highly focussed to a spot a few microns in diameter, or even less, and directed to specific locations on the sample. It is thereby possible to map the distribution of a range of elements across a sample. There are widespread applications of PIXE and PIGME techniques to geology (Brissaud *et al.*, 1986), archaeology (Duerden *et al.*, 1979) and environmental science.

7.5.3 *Thin layer activation*

If the energy of the proton beam is sufficiently high, interaction with the target elements will lead not only to the emission of X rays and gamma rays, but also to activation. Since the primary beam rapidly loses energy, the activation process is restricted to the region close to the surface and is technically known as thin layer activation. A number of industrial applications of this technology have been developed (IAEA, 1997, Longworth, 1998, Ch. 5). One of the most important is to the study of the rate of wear of machine components such as gear wheels. The principles of the method to effect thin layer activation are illustrated in Figure 7.11, showing the activation of the surface of a gear tooth. The gear was removed from a piece of machinery and the iron activated using a beam of 13.8 MeV protons from an accelerator, so generating ^{56}Co(^{56}Fe + p $\rightarrow$ ^{56}Co + n).

The radioisotope ^{56}Co has a half life of 77 d and emits gamma rays in the range 847 to 3253 keV. The cross section of the reaction is shown in Figure 7.11(a). The level of ^{56}Co activity generated depends on the number of protons incident on the target. This is expressed in terms of a yield Y which may be expressed in terms of kilobecquerel ^{56}Co per microampere hour. The yield of ^{56}Co decreases from about 500 kBq/mA h at the surface to virtually zero at the range of the 13.8 MeV protons in iron (say 190 μm). Following activation with the accelerator (cyclotron) beam, the machine is re-assembled, and one of two strategies is followed. The first involves circulating the lubricating oil through a filter designed to collect the metal particles which have been eroded off the component. The build up of radioactive particles on the filter may be monitored continuously (Figure 7.11(d)). The second approach involves locating the detector near the component and monitoring the rate of loss of the surface activated material. Since the monitoring is continuous, and does not interfere with the functioning of the machine, the effect of various factors such as the properties of the oil and its additives and the operating temperature of the machine may be investigated.

7.5.4 Smoke detectors

One of the most widespread applications of radioisotopes is in smoke detectors, millions of which are sold every year. A small ^{241}Am source is used in one of the common designs. Alpha particles from the source ionise air molecules which then migrate to a detector electrode to form a small ionisation current. In the presence of smoke particles, the current is reduced, and this triggers an alarm at a pre-determined level. Radioactive smoke alarms are sensitive to particles over a much wider size range than those based on light scattering.

7.6 Scientific and industrial applications of the absorption of radiation

7.6.1 The chemical effects of radiation

Industrial nucleonic gauges are designed to monitor key operating parameters in real time, to optimise process efficiency through on-line control and to diagnose plant malfunction. The absorption of ionising radiation in many materials also leads to chemical changes. However, exploitation of the chemical effects of radiation on an industrial scale requires radiation intensities that are thousands to millions times higher than those used for diagnostic purposes.

The absorption of high-energy radiation by a molecule, M, leads to a range of primary excitation and ionisation products (M* and M$^+$) which may be represented by expressions such as:

$$M \longrightarrow M^* \qquad \text{(excitation)} \tag{7.8}$$

$$M \longrightarrow M^+ + e^- \qquad \text{(ionisation)} \tag{7.9}$$

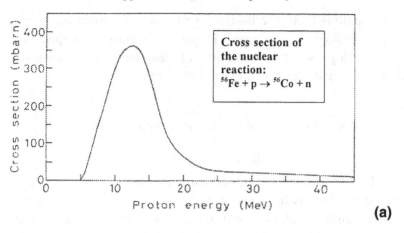

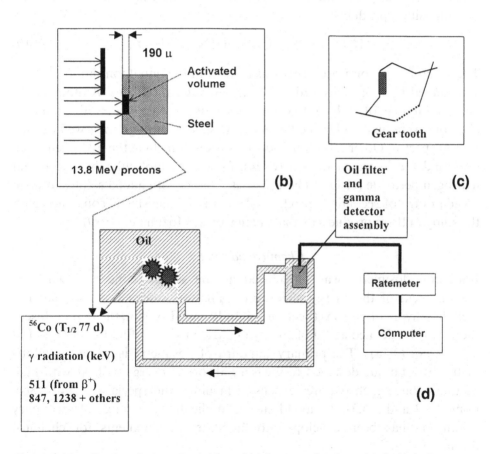

Figure 7.11.   Study of wear rate using thin layer activation. (a) Activiation cross section as a function of proton energy. (b) Irradiation of a gear tooth (shown in (c)) with 13.8 MeV protons. (d) Measurement of the rate of accumulation of wear debris on the filter (after IAEA, 1997, Figures 4 and 9).

Subsequent reactions form the basis of a number of industrially important chemical and biological processes. Important industrial applications of radiation technology are the formation of heat shrinkable polymers, the sterilisation of medical products and food irradiation.

7.6.2 Radiation chemistry of aqueous solutions

Basic processes

Since the mid-1920s it has been known that X rays as well as α particles could decompose water into hydrogen and oxygen. However, it was only in the 1960s that it could be shown that the initial ionisation and excitation processes (Eqs. (7.8) and (7.9)) lead to the following primary radiation decomposition products

$$H_2O \rightarrow e^-_{aq}, H^+, H\bullet, OH\bullet, H_2, H_2O_2. \tag{7.10}$$

The species e^-_{aq} comprises electrons formed in the initial ionisation process, surrounded by a partially ordered array of water molecules. Although they have a lifetime of only a few microseconds in pure water, the aquated electrons were finally identified as separate species through their well defined optical spectra. Other radiation induced products include the hydrogen atom ($H\bullet$) and the hydroxyl radical ($OH\bullet$), as well as molecular hydrogen and hydrogen peroxide (H_2O_2). These species interact with added solutes to form a wide range of secondary products. Radiation chemistry is concerned with the study of these products (Tabata *et al.*, 1991, Mozumber, 1999).

Chemical dosimetry

Radiation dosimetry was introduced as the science concerned with the measurement of the energy of ionising radiation absorbed in any material including living organisms (Section 2.4.2). In the SI system, the unit of dose is the gray (Gy) defined as 1 J of energy imparted to 1 kg of matter (1 Gy = 1 J/ kg, see Eq. (2.3a)). The primary standard of dosimetry is based on calorimetry, a technique designed to measure absorbed energy as heat with high accuracy, but which requires expensive facilities and specialised skills (Sections 2.3.2 and 5.4.3). To avoid such complications, a range of secondary dosimeters has been developed to facilitate measurements for chemical dosimetry.

Dosimeters are calibrated to measure the absorption of ionising radiation in materials of interest and are routinely used for that purpose (McLaughlin *et al.*, 1989). Important examples are listed in Table 7.9. Each dosimetry

Table 7.9. *Dosimetry systems (after McLaughlin, 1989).*

Dosimeter	Classification	Range*	Comments
Calorimetry	Primary dosimeters and reference dosimeters	0.2 to 2 Gy/s**	Radiation energy absorbed as heat in a thermally isolated mass and measured
Ionisation chambers	Reference and field dosimeters	300 to 20,000 Gy/s	Radiation induces secondary elecrons in the chamber walls and ionisation in the gas
Standard ferrous sulphate (Fricke dosimeter)	Chemical dosimeters, reference dosimeters	2 to 200 Gy	The radiation induced yield of ferric species from acidified ferrous sulphate is measured
Cerous ceric dosimeter	Chemical dosimeters, transfer dosimeter	10^3–5×10^4 Gy (up to 10^6 Gy possible)	The formation of ceric ions induced by the irradiation of acidified cerous or cerous/ceric solutions
Alanine dosimeter	Transfer dosimeter	1 to 10^5 Gy	Alanine mixed with paraffin, cellulose, polystyrene or other binders produces free radicals on irradiation which are measured by (ESR)
Plastic dosimeters	Routine dosimeters	Varies, but generally in the range 10^3 to 10^6 Gy	Changes in the optical absorption of a range of plastics or dyed polymers (e.g. red, amber or clear perspex). Colour changes in radiation sensitive films
Thermoluminescent (TLD) dosimeters	Routine dosimeters	refer comments	Usable dose range depends on the material: $Li_2B_4O_7$: 10^{-2} to 10^5 Gy CaF_2:Mn: 10^{-6} to 10^3 Gy $CaSO_4$:Dy: 10^{-1} to 5×10^6 Gy

* Calorimetry and ionisation chambers expressed in terms of the absorbed dose rate (Gy/s); chemical dosimeters expressed as total dose absorbed (Gy)

** A water/polystyrene calorimeter (US National Institute of Science and Technology) system has different properties and therefore different areas of applications, some of which are listed in the table.

Chemical dosimetry systems are important because they are widely used as transfer dosimeters in industry. A well known example is the ferrous sulphate or Fricke dosimeter. Its operation is based on measurements of the energy absorbed during ferrous–ferric oxidations, the absorbed dose being calculated from the readily measured concentration of ferric ions. Reliable dosimetry is fundamental to good radiation practice and underpins many of the applications introduced in the next section.

7.6.3 Industrial applications of high-energy radiation

Introductory comment

High-energy radiation is currently being applied to processes as diverse as the formation and degradation of polymers, the curing of paints and coatings, the sterilisation of medical products and the pasteurisation of certain foodstuffs and additives. A number of potentially exciting applications are under investigation. These include the pasteurisation of sewage and the treatment of flue gases from power generation. For a systematic discussion of these issues, readers are referred to the works of Bradley (1984) and Woods and Pikaev (1993).

Radiation induced polymerisation

The absorption of radiation in organic systems leads initially to ionisation and excitation and ultimately to the formation of free radicals. The organic free radical $R\bullet$ (Eq. (7.11)) is the analogue of the H atom and OH radical formed in aqueous solutions (Eq. (7.10)). The free radicals undergo the following reactions.

- Abstraction of atoms leading to the production of a new free radical, R', or radical centre on a polymer chain:

$$R\bullet + R'H \longrightarrow RH + R'\bullet \qquad (7.11)$$

- Addition to double bonds in olefin to form another radical:

$$
\begin{array}{c}
 H\ R'' H\ R' \\
R'\bullet + C{=}C \longrightarrow R'{-}C{-}C\bullet \\
 H\ H \phantom{\longrightarrow R'{-}C} H\ H
\end{array}
\left[
\begin{array}{c}
H\ R'' \phantom{\longrightarrow R'{-}C} H\ R''\ H\ R'' \\
+\ C{=}C \longrightarrow R'{-}C{-}C{-}C{-}C\bullet \\
H\ H \phantom{\longrightarrow R'{-}C} H\ H\ H\ H
\end{array}
\right] \qquad (7\text{-}12)
$$

This is an addition reaction which is known as a chain propagating step because it can be repeated many times, each time leading to a lengthening of the carbon backbone or chain shown in bold in Eq. (7.12). If the reagent monomer (CH_2=CHR″) were ethylene (CH_2=CH_2), the polymer product would be polyethylene with R″ replaced by H in Eq. (7.12). If the monomer were propylene (CH_3CH=CH_2), the plastic polypropylene would be formed.

Although high-energy radiation can be used to initiate polymerisation, it is not used on an industrial scale for the production of bulk polymers as it cannot compete economically with catalytic methods. It has, however, found specialist applications in the curing of surface coatings and the printing industry. Either electron beam irradiation or UV light is used. The latter is generally favoured as the capital costs of the installations are much lower.

Effects of high-energy radiation on polymers

Radiation effects are complex and may lead either to the enhancement of the polymer structure or to its degradation. In the former case, they result in the establishment of chemical bonds between adjacent polymer chains in a process known as cross-linking; in the latter, they lead to the degradation of the polymer chains through bond scission. Although both processes occur simultaneously, polymers may be classified according to whether cross-linking or scission dominates. Cross-linking polymers include polyethylene, polystyrene, PVC (polyvinyl chloride), and, in particular, heat shrink plastics. Examples of scission polymers are Teflon, cellulose and polypropylene.

(1) *Radiation cross-linking:* Cross-linking is the process by which adjacent polymer chains establish chemical bonds leading to complex networks. It results in the transformation of weak plastic materials into stronger, extensible, more rubber-like polymers. Free radical reactions are primarily involved. The absorption of ionising radiation by a polymeric material leads to the formation of a number of active sites along the chain. If two such sites on adjacent chains are in close proximity a chemical cross-linking bond will be formed.

The production of heat shrink plastics is a major industrial use of radiation cross-linking (Bradley, 1984, Ch. 10). It depends on the fact that lightly irradiated high-density polyethylene (but also other polymers) exhibit a so-called 'memory effect'. The concept may be illustrated with reference to the manufacturing steps.

Using normal manufacturing procedures, the polymer is formed into the required shape and then cross-linked with ionising radiation (electron beams or ^{60}Co radiation). The cross-linked polymer is heated above its melting point and is then blown or stretched into a larger version of the same shape. The larger version is then 'frozen' by cooling. When next heated, it reverts to the required size and shape.

Heat shrink material is manufactured as tubing, sheet or moulded components. The principal use of light walled tubing is in the electrical and electronic industries. Heavy walled tubing is used to protect splice connectors in telecommunication and power cables, and as a sleeve to protect oil and gas pipelines against corrosion. Heat shrink film is used for food wrap and other packaging.

(2) *Radiation induced degradation:* As indicated above, one of the effects of the irradiation of polymers is scission of the carbon chain leading ultimately to degradation. Although this process dominates in a range of polymers, it has found only limited industrial application. An exception is the degradation of Teflon.

Teflon or polytetrofluoroethylene (PTFE) is an analogue of polythene in which the hydrogen atoms are replaced by fluorine. Using doses of radiation of the order of 1 MGy (100 Mrad), Teflon can be sufficiently degraded to allow grinding into small particles. This has led to an economical method for the recycling of scrap Teflon into powders fine enough (~3 μm) for use in printing ink and as coating additives.

7.6.4 Radiation sterilisation

Introduction

Ionising radiations have long been known to be lethal to organisms at sufficiently high doses. It is not surprising, therefore, that a major industrial application of these radiations is radiation disinfection by means of sterilisation. Sterilising doses are of the order of 10 000 to 30 000 Gy (1 to 3 Mrad), delivered over a period of hours to days. This dose rate is a billion times greater and the total dose several million times greater than that received annually by the general public (Table 2.3).

Sterilisation of disposable medical products

This application has increased dramatically since the construction of the first commercial plants in the 1960s because sterilisation using ionising radiations has the following important advantages.

- In a well designed plant, the radiation reaches the whole of the product at a dose rate which can be controlled and accurately monitored; and
- Unlike other sterilisation processes, radiation technology is well adapted to continuous operation and can be undertaken at normal temperature and after final packing.

Three types of ionising radiation are used: ^{60}Co γ radiation, electron beam irradiation and bremsstrahlung from high-energy electron accelerators (3 to 6 MeV). The irradiated products include medical gloves, syringes and a range of pharmaceutical products.

Other applications

The potential application of radiation disinfection to the treatment of industrial waste water and sewage sludge has been demonstrated. Widespread introduction of the technology will depend on the production of reliable high-performance electron beam accelerators, which will allow a sufficiently high throughput of the waste stream, and on environmental considerations.

7.6.5 Food irradiation

One of the major problems associated with the provision of adequate nutrition to the world's population is the post-harvest loss of food, particularly in tropical countries. The potential use of radiation to increase the shelf life of foodstuffs has been studied for several decades. Nevertheless, there remains substantial resistance to the widespread introduction of irradiated foodstuffs due to community distrust of the technology and to economic considerations.

There is, however, firm scientific evidence that irradiation would not compromise the safety of food under properly controlled conditions (Diehl, 1990). Currently (in 1999) about 41 countries have cleared one or more food products for irradiation. The cumulative list comprises over 220 items. Comprehensive information is published on the Internet by the International Consultative Group on Food Irradiation at a site which is linked to the International Atomic Energy Agency web page (http://www.iaea.org).

Chapter 8

Application of tracer technology to industry and the environment

8.1 Introduction

8.1.1 Radiotracers come on the scene

The first use of radiotracers happened almost by accident. In 1912, two Austrian-Hungarian chemists G. de Hevesey and F.A. Paneth, research students at Professor Rutherford's Manchester laboratory, were challenged by Rutherford to demonstrate their chemical skills by separating minute traces of radium-D, as it was then known, from the lead with which it was associated and which had been derived from uranium deposits. They worked hard for two years but had to conclude that radium-D and lead must belong to a group of 'practically inseparable substances' later known as isotopes.

De Hevesey, a future Nobel Laureate, then conceived the idea of adding radium-D (subsequently identified as ^{210}Pb, a β particle emitting isotope of lead), to bulk lead and use the emitted β particles to trace sparingly soluble lead salts. It worked 'like a charm' and was the first example of radiotracing. Applications to many aspects of chemistry, biology and agriculture soon followed. Through the 1920s, progress was limited by the fact that the only radiotracers available were those separated from naturally radioactive materials. This was to change dramatically in the following years with:

- the discovery of artificial radioactivity by Frederick Joliot and Irene Joliot-Curie (early in 1933);
- the construction in 1932 of the first cyclotron designed to produce radionuclides by M. S. Livingston and E. O. Lawrence;
- the discovery of neutron activation by Enrico Fermi in the mid-1930s (Section 1.3.3); and

• the construction in 1942 of the Chicago pile (nuclear fission reactor), by a team also led by Enrico Fermi, which opened the doors for the large scale production of radiotracers (Section 1.3.3).

Tracers may be defined as substances which behave identically in essential respects to the selected component of a complex system and which can be measured independently of the other components and with high sensitivity. Many different types of tracers are available, but emphasis is here placed on the use of radioactive tracers. In developing a strategy for an investigation, the practitioner must select the key component of the system for labelling, chose the optimum radionuclide as a tracer and apply the most favourable injection procedure and measurement techniques.

A number of reviews on the application of tracer technology to industry and the environment have been published. The International Atomic Energy Agency (IAEA) has been in the forefront of promoting these applications. Readers are referred to the *Guidebook on Radioisotopes in Industry* (IAEA, 1990a) and the *Guidebook on Nuclear Techniques in Hydrology* (IAEA, 1983). A collection of essays on industrial radioactivity applications which has been found to be particularly helpful by many practitioners was edited by Charlton (1986).

8.1.2 Radiotracers: their advantages and their problems

The advantages of radiotracers, a summary

Tracer techniques are widely applied in engineering and environmental investigations because they allow the detailed study of individual components of complex systems. They are complementary to numerical modelling approaches which treat such systems as a whole, albeit with simplifying assumptions. A wide range of chemical and radioactive tracers is available. Concern is sometimes expressed in the workplace about the use of radioactive materials, and many decision makers tend to use non-radioactive alternatives if available. Despite this, radioactive tracing has proven resilient for the following reasons

• A wide variety of radionuclides (Tables 8.1 and 8.2) and labelled compounds are commercially available, making it likely that a suitable tracer will be available for the task in hand.
• Radioisotopes can be measured accurately with high sensitivity at high dilution.
• Radiotracers are usually monitored external to the plant and so do not interfere with production schedules.

Table 8.1. *Gamma ray emitters extensively employed for applications. (Decay data from* Charts of Nuclides *published since 1985.)* $E(_{KX})_{av}$ *and* E_γ *are in kiloelectronvolts;* f_{KX} *and* f_γ *are expressed as percentages.*

Nuclide	Decay mode	$T_{1/2}$	$E(_{KX})_{av}$ (f_{KX})	$E_\gamma(f_\gamma)$
^{7}Be	EC	53.28 d		478(10.6)*
^{11}C	β^+	20.38 m		511(200)*, from β^+
^{18}F	β^+	109.7 m		511(194)*, from β^+
^{22}Na	β^+	2.602 y		511(180), from β^+, 1275(100)*
^{24}Na	β^-	14.96 h		1369(100), 2754(100)*
^{46}Sc	β^-	83.3 d		889(100), 1121(100)*
^{47}Sc	β^-	3.351 d		159(68)*
^{51}Cr	EC	27.70 d		320(9.85)*
^{52}Mn	EC	5.59 d		744(90), 935(95), 1434(100)
^{54}Mn	EC	312.2 d		835(100)*
^{56}Co	EC	77.49 d		847(100), 1038(14), 1238(67), 1771(16), 2598(17) plus many others
^{57}Co	EC	271.7 d		122(86), 136(11)
^{58}Co	β^+	70.82 d		811(100), 511(30)*, from β^+
^{59}Fe	β^-	44.51 d		1099(56), 1292(44)
^{60}Co	β^-	5.27 y		1173(100), 1332(100)*
^{65}Zn	EC	243.9 d		1116(51)*
^{67}Ga	EC	3.26 d		93(38), 185(21), 300(17)
^{75}Se	EC	119.8 d		121(17), 136(59), 265(59), 280(25), 401(11)
^{82}Br	β^-	35.34 h		554(71), 619(43), 698(29), 777(84), 828(24), 1044(27)
^{85}Sr	EC	64.85 d		514(98)*
^{88}Y	EC	106.61 d		898(94), 1836(99)* (Figure 6.6(a))
^{95}Zr	EC	64.00 d		724(44), 757(55)*
^{99}Mo	β^-	65.92 h		141(91), 740(14), plus many others
^{99m}Tc	IT	6.007 h		141(89)* also in equilibrium with Mo-99, see also Figure 4.7
^{109}Cd	EC	462.6 d	23(102)	88(3.7)*
^{110m}Ag	IT	249.8 d		658(94), 678(11), 707(17), 764(22), 885(73), 937(34), 1384(24), 1505(13), plus many others
^{124}Sb	β^-	60.20 d		603(98), 723(11), 1691(48)
^{125}Sb	β^-	2.76 y	28(46)	428(29), 463(10), 6 00(18), 636(11)
^{131}I	β^-	8.021 d		364(81)
^{133}Ba	EC	59.6 d	33(118)	79/81(37), 356(62) (Figure 3.9)
^{134}Cs	β^-	2.065 y		569(15), 605(98), 796(85)
^{137m}Ba	IT	2.55 m		662(85)*, from Cs-137
^{139}Ce	EC	137.6 d	35(80)	166(79)*
^{140}Ba	β^-	12.76 d		537(24)*, (Section 1.6.3)
^{140}La	β^-	40.28 h		329(21), 487(46), 816(24), 1596(95) in equilibrium with Ba-140
^{141}Ce	β^-	32.50 d	37(16)	145(49)*

^{144}Ce	β^-	285.0 d		134(11) followed by Pr-144, $T_{1/2} = 17$ m, β^- emitter
^{152}Eu	β^-/EC	13.50 y	42(74)	122(28), 344(27), 779(13), 964(15), 1112(14), 1408(21) plus many others (Figure 3.14(a))
^{153}Gd	EC	241.6 d	43(120)	97(30), 103(21)
^{154}Eu	β^-	8.55 y	44(25)	123(40), 723(21), 873(12), 996(11), 1005(18), 1275(35) plus many others
^{169}Yb	EC	32.03 d	54(185)	63(44), 110(17), 131(11), 177(22), 198(35), 308(11) plus many others
^{182}Ta	β^-	114.4 d		1121(35), 1189(16), 1221(27), 1231(12), plus many others
^{192}Ir	β^-/EC	73.83 d		296(29), 308(30), 316(83), 468(48)
^{197}Hg	EC	64.1 h	71(72)	77.3(18)*
^{198}Au	β^-	2.696 d		412(96), (Figure 3.15)
^{203}Hg	β^-	46.60 d	74(13)	279(82).*
^{207}Bi	EC	32.2 y	78(68)	570(98), 1064(74)
^{241}Am	alpha	433 y	LX17(38)	59.5(36)*, (Figure 5.7)

Notes: Only the more energetic and abundant photons are listed, though there are a few exceptions. X rays are shown if $(E_{KX})_{av} > 20$ keV and $f_{KX} > 10\%$, gamma rays if $E_\gamma > 50$ keV, $f_\gamma > 10\%$, all rounded to the nearest whole number and the nearest percent with the percentage shown in brackets. The entry 'plus many others' denotes the emission of large numbers, upwards to 50 and more, of very low-intensity γ rays which are unlikely to affect measurements. An asterisk (*) denotes that no other γ rays were emitted except possibly with very low intensity. Uncertainties in the quoted data are well within ±1.0%.

Portable sources should always be convenient to use but completely safe against contamination (see Section 4.2.3).

These factors apply in different ways to different applications. A helpful discussion can be found in Charlton (1986, Ch. 7 by T. L. Jones, 'Planning a radioisotope tracer investigation').

Radiation safety

The handling of radioactive materials is highly regulated to ensure that doses to radiation workers are readily kept well below the recommended limits (Figure 2.3). This is achieved by providing adequate shielding, minimising the time of potential exposure and maximising the distance between the operator and the source (Section 2.6). Good experimental design ensures that doses to operators are not only well below officially regulated guidelines, but, in addition, are as low as reasonably achievable (the ALARA Principle, Section 2.5.1). Under the current regulatory environment, occupational health risks to radiation workers are at least as low as, and frequently much lower than, those accepted by workers in industries not using ionising radiation.

Table 8.2. *Pure or nearly pure β particle emitters supplied as sealed sources for transmission or backscatter measurements.*[a]

Nuclide	$T_{1/2}$[b]	E_β(keV) max	E_β(keV) ave[c]	Average range (mg/cm²Al[d])	Per cent[e]
³H[f]	12.4 y	18.6	6.7		100
¹⁴C	5730 y	156	50	5.5	100
⁶³Ni	100.1 y	67			100
⁸⁵Kr[g]	10.73 y	672	251	80	100
⁹⁰Sr –	28.3 y	545	195	55	100
⁹⁰Y[h]	(64.1 h)	2270	940	500	100
¹⁰⁶Ru –	369 d	39	10	–	100
¹⁰⁶Rh	(30.4 s)	3550	1480	850	79
¹⁴⁴Ce –	284.3 d	316	91	–	77
¹⁴⁴Pr	(17 m)	3000	1230	710	98
¹⁴⁷Pm	2.62 y	225	62	8.0	100
²⁰⁴Tl	3.78 y	763	245	80	98
²¹⁰Pb –	22.3 y	15	5	–	100
²⁸⁰Bi	(5.01 d)	1160	390	160	100

[a] See Sections 3.3.3 and 4.2.3 for comments on sealed sources of α particle emitters.

[b] The parent–daughter pairs should be in equilibrium when the half life of the parent applies to both (Section 1.6.2). The daughter half life is added in brackets.

[c] This column lists the average energies of the β particle spectrum (see Section 3.3.4).

[d] See Figure 3.2(b) about ranges of electrons in aluminium. When engaged on applications the range of electrons in the material of interest can be verified the more accurately the nearer it is to the mean values of the spectrum listed in this column.

[e] Without going into details, several of the decays are subject to small percentages (<2.5%) of γ rays (and there are always the background γ rays), and a few low-energy β branches. Effects due to these 'impurities' can normally be ignored, but care is called for.

[f] ³H (tritium) α particles are useful for low-energy bremsstrahlung sources but rarely for thickness measurements.

[g] Sealed krypton sources are supplied as small tubes each fitted with a thin window.

[h] For the parent–daughter pairs the β particles of interest are emitted by the daughter.

Rules-of-thumb for calculating absorbed whole body doses due to γ rays which are of the order of the natural background dose are given in Section 2.7.2, while Table 7.3 lists half value layers (HVLs) for aluminium, iron and lead for γ ray energies in the range 0.2 to 1.5 MeV. For detailed information on radiation safety, readers are referred to the textbooks cited in Appendix 2.

8.1.3 The evolution of radiotracer applications

Early examples of tracer applications

Prior to World War 2, applications of radioisotope tracing were on a relatively small scale and restricted to the fields of chemistry, biology and agriculture. In the immediate post-war period, there was considerable interest in extending the techniques to industry. Progress was rapid and was reported in the proceedings of two famous international conferences on the Peaceful Uses of Atomic Energy sponsored by the United Nations (UN, 1956, 1958). They became known as the 1955 and 1958 Atoms for Peace conferences. A few examples of early applications from the petroleum sector are now described.

(1) *Oil field studies:* Not surprisingly, radioisotopes were first applied to those industries in which chemical tracers were already well established. For instance, in oil field engineering, the injection of water is widely used to increase the yield of valuable hydrocarbons extracted at the production wells. Information on the reservoir conditions near such wells can be obtained from drilling logs and other tests. However, tracers are needed to obtain critical information on flow patterns in the oil-bearing strata between the injection and the production wells. Fluor-escein and other dyes, as well as caesium and selenium were traditionally used, but were rarely satisfactory because of excessive adsorption on the solid surfaces. Early on, Watkins and Dunning (UN, 1956) reported that radiotracers including I-131 and Ir-192 in a suitable chemical form (Table 8.1) were much more satisfactory.

(2) *The role of tracers in the oil refining industry:* Applications of radioisotope techniques to oil refining have been undertaken since the early 1950s. Hull and Fries (UN, 1956) reviewed early studies which included much of their own work. Radiotracer techniques were developed during that period for

the location of leaks in pipelines
the measurement of flow under a wide range of conditions, and
the study of the behaviour of catalyst in refinery columns.

Between the first and second Atoms for Peace Conferences (1955 and 1958), applications were extended to virtually every industrial sector. The IAEA, which was founded in 1958, has played a major role in extending and promoting the applications of tracer technology, particularly to developing countries.

Recent advances

Since the early 1960s, major progress has been made in the fields of micro-electronics, data processing and visualisation, and the numerical modelling of complex dynamic processes.

Although not directly related to the science of radiation and radioisotopes, these developments have had a major effect on the scope and usefulness of tracer technology. Many of the most recent advances were developed for the health sciences and later transferred to industry. Examples include industrial computerised tomography (Section 7.2.1) (Wellington and Vinegar, 1987, Romans, 1995), the industrial applications of gamma ray camera systems (Castellino *et al.*, 1984, Jonkers *et al.*, 1990) and of SPECT (Single Photon Emission Computed Tomography) (Bergen *et al.*, 1989).

The following sections will deal with the design of radiotracer investigations, including the acquisition of precise data. Applications to studies of fluid flow, residence time distributions and contaminant dispersion will be discussed with reference to case studies likely to be of wide interest. Table 8.3 lists a number of industrial and environmental applications of artificial tracers.

8.2 Tracer applications in the field

8.2.1 The general concept of the radiotracer experiment

The principle of the methods underlying most radiotracer experiments is illustrated in Figure 8.1. A small quantity of a radiotracer is injected at point A, disperses through the system and is monitored at B. The concept of a 'system' should be interpreted in the broadest sense. A system might comprise, for example:

- liquids or gases flowing in industrial pipelines
- water flowing in rivers or streams
- reagents mixing and reacting in petrochemical or mineral processing plant
- contaminants dispersing in rivers or estuaries opening to the ocean
- sand and sediment moving on the sea bed.

The output from such a system is either the countrate from a strategically located detector or the radioactivity concentration of the tracer in samples of the fluid. Fifty or more detectors may be deployed to study the dispersion of tracers in complex investigations. The practitioner is challenged to extract the best quality information from the available data. Readers are referred to the *IAEA Guidebook on Radioisotope Tracers in Industry* for further details (IAEA, 1990a, pp. 39–92).

Table 8.3. *Applications of artificial tracers to industry*
(see Tables 8.1 and 8.2 for decay data).

Industry sector	Investigation	Applications	Comments
Refining	Fluidised catalytic cracking units (FCCUs)	Velocities of vapour and the catalyst in the riser; separation efficiency of the catalyst and vapour	See Section 8.4.2
	Leakage and blockages in pipelines and chemical reactors	* Short circuiting through catalyst * leakage and blockages in sub-surface pipelines	
	Fluid flow rates	* Meter calibration – liquid and gas flow * Mass balance studies	See Section 8.3
Oil and gas field	Effectiveness of enhanced oil recovery strategies based on injection of water (with additives) or gas into the field at selected locations	* Used to identify, for example, short circuiting; * Validation of mathematical models of the process	Tracers are added at the injection wells and sampled at the production wells. The tracers include HTO (or $^{60}Co(CN)_6{}^{3-}$) for water and ^{85}Kr or T (or ^{14}C) labelled methane for the gas
	Leakage between gas bearing strata in a production field	^{85}Kr or T (or ^{14}C) labelled light hydrocarbons injected at observation wells and migrate to production wells	
Chemical	Flow rate measurements for *in situ* meter calibration	Point to point and tracer dilution methods widely used	See Section 8.3.2
	Mercury inventories	Tracer dilution using ^{203}Hg and/or ^{197}Hg	See Section 8.4.4
	Leakages and blockages		Refer to entry under Refining above
Minerals	Process optimisation	Frequently involve (RTD) studies	See Sections 8.3.7 to 8.3.9 Widespread application to the minerals processing industries, e.g. gold

Table 8.3 (*contd*)

Industry sector	Investigation	Applications	Comments
	Hot processing of ores and concentrates	Wide range of tracer techniques to study the processes in blast furnaces and roasting furnaces	Refer to IAEA (1990a, Ch. 6)
	Electrorefining	Measuring the efficiency of the electrolytic production of Al	Current efficiencies of 0.2% possible if all sources of error rigorously controlled. ^{198}Au used as there is no suitable Al tracer
Iron and steel	Blast furnaces	Refractory lining wear. Sealed ^{60}Co sources into selected refactory bricks during lining – as bricks erode pellet also erodes and radiation level falls	Techniques also available to detect oxide impurities in steel resulting from the erosion of refractories
	Blast furnaces	Simultaneous RTD studies on the iron and slag	See Section 8.4.3
Cement	Pre-mixing silos	Efficiency of mixing of the four components, stock in silos	Similar applications in the paper industry
	Rotary kiln	Residence time studies	
Mechanical engineering	Wear tests	Activation of pistons or cylinders and monitoring of activity of eroded material in test rig	Thin layer activation (TLA) often used (Section 7.5.3)

8.2.2 Choice of the optimum radiotracer: general considerations

Introduction

When selecting a radiotracer from those listed, for example, in Tables 8.1 and 8.2 or from collections of decay data listed in Section 4.2.1, the following requirements have to be balanced.

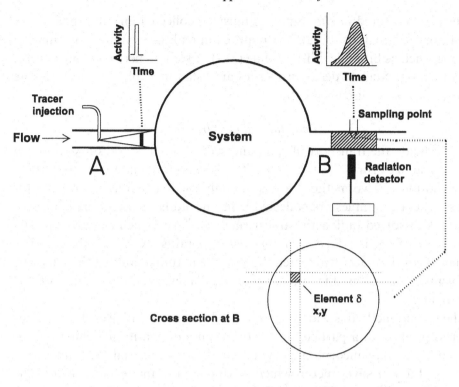

Figure 8.1.   The principle of a tracer investigation. Radiotracers are used to track individual components of complex systems. A tracer is injected at A and its response monitored at B and interpreted in terms of the information sought on the behaviour of the system. The location of the radiation detector and the sampling point δ is shown.

- The radiotracer should behave identically in essential respects to the component of the system which is under study.
- Consistent with the aims of the investigation, a shorter rather than a longer lived radionuclide should be chosen.
- If practical, it is preferable to choose radionuclides emitting γ rays in the range 150 to about 500 keV; at higher energies there are greater problems in handling and transport.

The process of choosing an appropriate radiotracer will be illustrated with examples from environmental and industrial applications.

Water tracing

On the face of it, the use of HTO (tritiated water) as a tracer for water or aqueous systems would appear to be an obvious choice. However, tritium is a very low-energy β particle emitter ($E_{max} = 18.6$ keV) and hence its activity cannot be measured *in situ*. Samples must be collected and returned to the laboratory for analysis. If data are required in real time, gamma ray emitting isotopes such as bromide (^{82}Br), iodide (^{131}I) or sodium pertechnetate (^{99m}Tc) must be used. Nuclear decay properties are listed in Table 8.1 and flow rate measurements are discussed in Section 8.3.

Sand and sediment tracing

To investigate the transport of sand and sediments, it is necessary to match the particle size distributions and density of the natural material with those of the radiotracer. Two methods are commonly used. The first is preferred for fine sediments. Samples are collected from the study area and made up as a slurry. A dissolved radioactive tracer such as ^{198}Au, which adsorbs strongly onto the surface, is added in a shielded apparatus. Although the physical properties of the tracer will match those of the natural material, there may be problems in assuring the integrity of the label over the period of the investigation.

The second method is widely used in sand tracing. It involves the synthesis of glass beads with a particle size distribution and density matching that of the sand. The glass incorporates the required target material (e.g. lanthanum oxide, iridium or silver metal) which is irradiated in the reactor to form the tracer (^{140}La, ^{192}Ir or ^{110m}Ag, see Table 8.1). The integrity of the label is not in question. However, there may be problems in matching the particle size distributions.

In both cases the radiotracer is released to the area under investigation and its behaviour monitored by means of the emitted radiations. As noted earlier, the half life of the isotope should be as short as possible consistent with the aims of the investigation. For instance, ^{198}Au ($T_{1/2} = 2.70$ d), is commonly used for studies which take up to a week; ^{192}Ir ($T_{1/2} = 73.8$ d) for one or two seasons; and ^{110m}Ag ($T_{1/2} = 249.8$ d) for investigations which extend over a complete annual cycle.

Industrial tracing

Industrial applications provide numerous examples of the need to optimise the choice of radiotracers. In the oil refining industry, for instance, it is frequently necessary to employ tracers in chemical forms which are soluble in hydrocarbons but will not be extracted from the hydrocarbon into the aqueous phase.

In the study of blast furnaces (Section 8.4.3), the metallic tracers gold-198 and cobalt-60 have long been used to trace the iron, and lanthanum oxide (^{140}La) to simultaneously trace the slag which is a mixture of oxides. Other investigations require gas tracers and ^{85}Kr (Table 8.2), a chemically inert noble gas, is commonly used. In some applications ^{41}Ar, another noble gas, is preferred as it has a short half life (1.83 h) and emits 1.3 MeV γ rays (100%). However, its short half life restricts applications to sites sufficiently close to a nuclear reactor.

8.2.3 Isotope injections

The isotope injection system should be designed to minimise the dose to personnel while maximising the amount of information gained from the investigation. Very short duration (instantaneous) injections are normally preferred because the experimental arrangement and the subsequent inter-pretation of the measured data are relatively simple. Occasionally, there is a need to inject a tracer into the system at a constant rate over the entire duration of the investigation. Examples of both these procedures will be presented later in this chapter.

8.2.4 Tracer detection and monitoring in the field

Field monitoring systems

These systems comprise radiation detectors, ratemeters and either data loggers or output devices. When working with X or γ rays, NaI(Tl) crystals optically coupled to a photo-multiplier tube and associated electronics are almost universally used as radiation detectors because they are relatively cheap, efficient and robust. Either 50×50 mm or 25×25 mm crystals are usually used. There is, as a rule, little incentive to use larger crystals because, as described below, the bulk of the signal is scattered radiation with energies below 150 keV. At these energies, the absorption efficiencies approach 90% even when using 25×25 mm crystals (Figure 4.4).

The radiation detector (Figure 8.2) is mounted in a cylindrical aluminium or stainless steel case. Although more rugged, a stainless steel housing absorbs a higher proportion of the low-energy radiation than aluminium. The loss of efficiency could exceed 50% and practitioners may thus need to chose between enhanced efficiency and reduced ruggedness.

The instrumentation required for routine work with NaI(Tl) crystals was described in Section 4.3.1. With field equipment, the electronics are usually stand-alone units including amplifiers, discriminators and ratemeters oper-ated by a battery. The data are frequently stored for 'on-line' or later transfer to computers or other output devices.

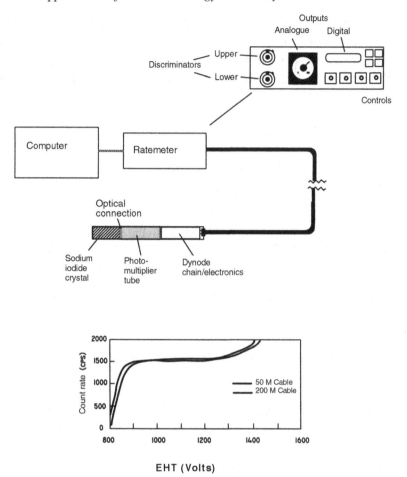

Figure 8.2.　Schematic representation of field monitoring equipment. This system comprises a field detector with a Na(Tl) detector, photo-multiplier tube and associated electronics. The amplified pulse is fed into the ratemeter which may be controlled by a computer. The ratemeter has an upper and lower discriminator and an analogue and digital output. The variation of the countrate response to the applied voltage which determines the counting plateau is shown.

The role of scattered radiation in the monitoring of radiotracers

One of the challenges of field investigations is to obtain the required information with the minimum quantity of radioactive material. Careful thought must be given to all aspects of the investigation from the conceptual design to the efficiency of detection of the radiotracer.

In many field investigations, the detection efficiency of a gamma emitting tracer is enhanced by counting the low-energy Compton scattered radiation.

The scattering mechanism leads to a significant build-up in the number of detectable γ rays, as shown in Figure 4.6(b) and discussed in Section 4.4.4. There is consequently a net gain in the sensitivity of the counting system to the tracer (Section 4.4.1). However, there is a price to pay. Spectral information that can only be obtained from an analysis of countrates in the full energy peak is largely lost. This is not important if, as is often the case, only a single radiotracer is involved.

The measurement of scattered radiation involves integral counting techniques. The lower level discriminator is adjusted to accept as much of the low-energy radiation as possible while avoiding noise pulses which could become significant when detecting pulse heights below a few millivolts. The setting of the upper level discriminator is adjusted to maximise the signal to background ratio. Readers are referred to Section 4.4.2 and Figure 4.7 for further information.

System checks and calibrations should be undertaken under conditions as close as possible to those in the field. The location of the counting plateau, and hence the operating voltage (Figure 8.2) depends on the characteristics of the detector, and also on the energy of the absorbed radiation. It is therefore common practice to establish the plateau position with an americium-241 source as its 59.5 keV γ ray is comparable in energy to much of the scattered radiation. Another option is to place a sample of the radiotracer in a thin lead pot and set the plateau using the degraded radiation emitted from the surface.

Accurate field measurements

Radioisotope investigations are becoming increasingly sophisticated, requiring good quantitative data. Procedures for accurate activity measurements in laboratories were discussed in Sections 6.3 and 6.4. The accuracy of field measurements depends on the quality of the calibration of the detector and this is normally limited by the need to reproduce in the laboratory the counting geometry which is found in the field.

Three classes of counting geometry may be distinguished:

- quasi-infinite geometry, when the detector is completely immersed in a large volume of liquid such as industrial tanks, rivers or lakes;
- quasi-planar geometry, when the detector is mounted, for instance, on a sled a few centimetres above labelled sediment on the bed of an estuary; and
- quasi-linear geometry when the detector is mounted external to a pipeline.

These are limiting cases which are discussed in the following chapter and illustrated in Figures 9.2(a) to 9.2(c) respectively.

Table 8.4. *Fluid dynamic investigations based on tracer dilution and mixing.*

Investigation	Measurement	Mixing conditions	Comments
Dispersion	Dispersivities and/or dispersion coefficients	Not applicable	Concentration profiles monitored as a function of distance from point of injection (Section 9.3.3)
Flow rate measurements	Volume flow and/or mass flow	Complete mixing at the measurement cross section	Sections 8.3.2 to 8.3.6
Residence time studies	Residence time distribution (RTD)	Complete mixing at the inlet and the outlet of the vessel	Readily achieved if inlet and outlet comprise narrow pipelines (Sections 8.3.7 to 8.3.9)

8.3 Applications of tracer technology to flow studies

8.3.1 General principles

Introduction

A general description of the application of tracers to scientific and engineering investigations was given in Section 8.2.1. Fluid dynamics, to be introduced in this section, has been one of the most fruitful fields of application. Three classes of investigations are listed in Table 8.4.

Residence time distribution (RTD)

The concept of the residence or transit time of a fluid particle which is being transported through a dynamic system (for instance, a chemical reactor, a section of pipeline or the reach of a river) is fundamental to many industrial and environmental applications. In particular, the residence time distribution (RTD), which is defined below, is often used to specify the performance of vessels designed for efficient chemical or bio-chemical engineering processing. Tracer techniques have been developed to measure the RTD of such vessels and to confirm their specifications (Sections 8.3.7 to 9).

Let us assume that a fluid flows through a system from an inlet (A) to an exit pipeline (B) (Figure 8.1). If a radiotracer is injected at A as a point source, the subsequent response of a detector at B will be a dispersed pulse since no two particles of fluid have identical transit or residence times. The distribution of transit times of all particles comprising the bulk fluid flowing through the tank is described as the RTD. It will be shown in Eq. (A4.3) of

Appendix 4, that the countrate response of the detector at B is a direct measure of the RTD of the tracer (and fluid) particles provided that

the tracer is injected as an instantaneous pulse, and
the bulk flow rate through the tank is constant

This simple result is the basis of many radiotracing applications to the chemical, refining and minerals processing industries where the efficiency of a process depends on the contact times between different reagents (Section 8.3.9).

Mean residence time (MRT)

The mean residence time (MRT) is the average time taken by the tracer (and fluid) particles to travel between the injection and measurement points. If there is an instantaneous injection at A, and a sharp response at B (Figure 8.1), the MRT is the time interval between the two. In more complex situations, the MRT is calculated from an analysis of pulse shapes as noted in Section 8.3.3 and in Appendix 4.

Complete mixing

A pulse of activity injected into a pipeline or a river is transported by advection (i.e. it moves with the fluid flow) and simultaneously disperses in three dimensions (Section 9.3.3). The effect of flow boundaries such as the walls of a pipeline or the banks of a river is to restrict the scale of turbulence and to induce mixing. The degree of mixing increases with distance from the injection point. As discussed below, the concept of complete mixing of the tracer with the bulk flow is extremely important. Complete mixing is achieved when the radiotracer concentration is everywhere the same following a continuous injection at a constant rate into a constant flow. However, pulse injections are usually favoured over continuous tracer injections because they generate the required information with the use of much less radiotracer. Under these conditions, tracer concentrations are nowhere constant, and the criterion for complete mixing may be expressed in one of the following ways.

(1) If the fluid is sampled at a constant rate (location B, Figure 8.1), complete mixing is achieved if the total amount of tracer removed during the passage of the entire radioactive pulse is independent of the sampling point (element δ) on the measurement cross section.

(2) If a detector is located within the fluid, complete mixing is achieved if the total number of counts (corrected for background and detector efficiency) registered during the passage of the plume is independent of its location (Figure 8.1) on the measurement cross section.

Further details are presented in Appendix 4. Tracer dilution techniques can be used to determine flow rates in a simple and direct manner provided complete mixing has been achieved. Herein lies the significance of the concept. It follows that the related concept of 'mixing length' or the distance from the point of injection until mixing is essentially complete is of considerable practical importance. A number of formulae have been developed to determine mixing lengths, some of which are listed in Table 9.3.

8.3.2 *Flow rate measurements: an overview*

There are two fundamentally different approaches to measuring flow rate: the pulse velocity ('point to point') method and tracer dilution techniques. The tracer dilution techniques may themselves be sub-divided according to the nature of the injection and of the measurement methods as shown below

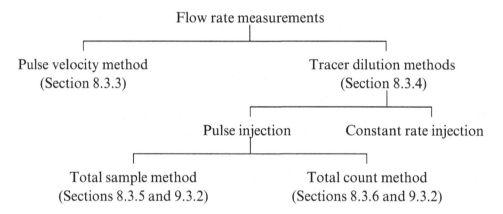

Flow rates may be obtained using tracer techniques without interfering with plant operation. Further, the fluid may be corrosive, and may be in the liquid or gaseous state or even multi phase. However a disadvantage is that the measurements are specialised and expensive and refer only to the conditions prevailing during the passage of the pulse past the monitoring points. Tracer techniques are therefore mainly used to calibrate installed metering systems designed for continuous readings.

8.3.3 *Flow rate measurements: transit time techniques*

Pulse velocity method

Conceptually, the pulse velocity method is the simplest and most widely used technique. A radioisotope pulse is injected into a system at X and its rate of

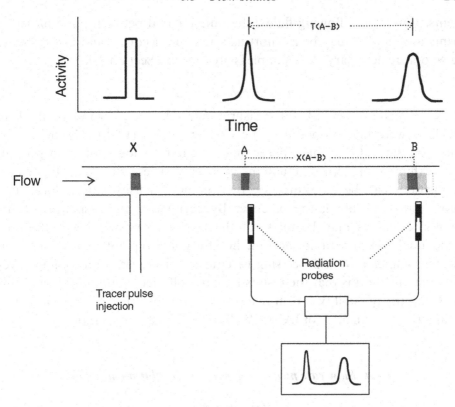

Figure 8.3. Flow rate measurements: the pulse velocity or 'point to point' method. The tracer pulse is monitored at two locations following an instantaneous injection. Normally, the pulse maxima are adequate to define the interval T_{A-B}. For the most accurate work, the pulse profiles must be analysed.

traverse between two appropriately selected points, A and B, is measured with externally located detectors (Figure 8.3).

The average velocity of the pulse, V_{A-B}, is given by

$$V_{A-B} = x_{B-A}/T_{B-A} \qquad (8.1)$$

where x_{B-A} is the distance between detectors located respectively at A and B, and T_{B-A} is the corresponding transit time. Normally, the interval between the pulse maxima is an adequate measure of the transit time for the passage of the pulse. However, for greatest precision, T_A and T_B are individually calculated from the shape of the response curves under conditions of complete mixing (Appendix 4, Eq. (A4.5)). To calculate the volume flow Q_{A-B} from the linear velocity V_{A-B}, it is necessary to know the cross sectional area of the pipeline.

The method is extensively used for both liquid and gas velocity measurements. For rapidly flowing fluids, the time interval between pulse measurements at A and B may be as short as a few milliseconds, and fast response detectors are necessary. An example is presented in Section 8.4.2.

Correlation methods

Cross correlation methods for measuring the velocity of two-phase fluids in pipelines were first proposed by Le Guennec *et al.* (1978). The method is based on the fact that multi phase fluids are not homogeneous and periodically exhibit short-term fluctuations in density. These characteristic density fluctuations can be monitored by two or more density gauges located at known intervals along the pipeline. By correlating the responses of the gauges, estimates may be made of the transit times and hence the flow velocities. Unlike the radiotracer method for point to point velocity measurements, significant data processing is required. However, an advantage over the tracer method is that the results may be collected continuously, and this opens up the possibility for on-line process control. Readers are referred to Chapter 5 (Section 5.12) of IAEA (1990a) for further information.

8.3.4 Flow rate measurements: tracer dilution methods

Introduction

Tracer dilution methods are widely used to measure the flow rates of liquids and gases. In contrast to the peak to peak method, volume flow rates may be calculated without knowledge of the cross sectional area of flow provided complete mixing has been achieved. Dilution methods are therefore well suited to measurements of the flow rates of rivers (Section 9.3.2). The tracer may be injected at a constant rate or as a pulse. Two variants of the pulse injection method, namely the total sample and total count methods, are introduced below and discussed in Appendix 4.

Tracer injection at a constant rate

Tracer of specific activity C_0 (Bq/l) is injected at a constant flow rate q (l/s) into a pipeline or stream in which the flow rate is Q (l/s). After complete mixing has been achieved, the diluted tracer is sampled and the activity c (Bq/l) measured (Figure 8.4). Since the mass, i.e. the activity of the tracer must be conserved, one has:

$$qC_0 = (Q + q)c. \tag{8.2}$$

Normally $Q \gg q$, and hence the volume flow rate is given by:

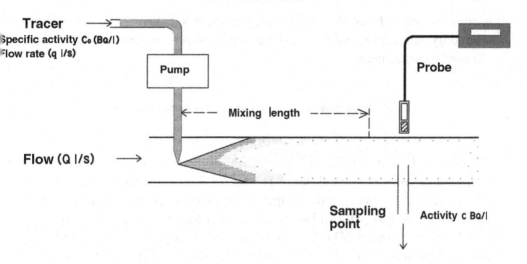

Figure 8.4.   Flow rate measurements: the continuous dilution method. The tracer is injected at a constant rate and monitored after complete mixing has been achieved.

$$Q = qC_0/c. \tag{8.3}$$

Experimentally, it is only necessary to measure q and the activity ratio C_0/c. Absolute activities and therefore absolute detector efficiencies are not required. For the most precise measurements, the activity ratios should be close to unity. Significant problems arise when the activity ratios exceed 10. Readers are referred to the discussions of the total count method in Section 8.3.6 and the measurement of inventories in Section 8.4.4.

As with other tracer dilution methods, measurements at constant injection rates are independent of the cross sectional area of the pipeline, assuming complete mixing. However, as noted earlier, the total amount of tracer is considerably greater than that required for the pulse dilution method described below.

A detailed description of the tracer dilution method can be found in Charlton (1986, Ch. 8). An example is provided of an industrial application of the constant injection method. The aim of the study was to provide an independent assessment of the accuracy of the dilution method as a basis for calibrating orifice plate meters used for measuring the flow rates of organic liquids. For the purposes of the test, the liquid was metered from the stock tank to a road tanker on a weigh bridge. The tracer was a completely miscible Br-82 labelled (Table 8.1) par-dibromo benzene. Careful checks were made of the constancy of the injection rate and of the stability of the counting system. In the end it was found that the tracer dilution and the weighing method agreed to better than 1%, which is completely satisfactory as a basis for secondary calibration and is a measure of what can be achieved with carefully planned measurements.

8.3.5 *Flow rate measurements: total sample method*

Principle of the method

As indicated above, pulse dilution methods may be classified into total sample and total count methods. In the total sample method a tracer of activity A (Bq) is injected into a pipeline or a river at position A (Figure 8.1) as an instantaneous pulse and sampled at a constant accurately measured rate q (l/s). The radiotracer HTO is commonly used for aqueous systems and ^{85}Kr for gas flow measurements. The sample is collected over an interval which includes the passage of the whole pulse. The tracer activity in the sample is measured (a Bq), and compared with the total injected activity. Assuming complete mixing had been achieved at B, the total flow rate Q (l/s) is given by the simple expression (Appendix 4, Eq. (A4.9))

$$Q = qA/a. \tag{8.4}$$

Again, it is not necessary to measure absolute activities but only the activity ratio. This considerably simplifies the measurements as the absolute efficiencies of the detectors are not required. The measurement of river flows using the total sample method is outlined in Section 9.3.2. Gas flow rate measurements will be discussed below.

Case study: gas flow rate measurement

The principle of the total sample method to the measurement of gas flow rates is illustrated in Figure 8.5. An accurately measured aliquot of tracer ^{85}Kr of activity A (Bq) from a stock supply is injected into the pipeline as a pulse (Figure 8.5(a)) and sampled for a known time ΔT after complete mixing was achieved (Figure 8.5(b)). As shown in the figure, the gas is collected into two bags. The bags are isolated from the pipeline, their contents are thoroughly mixed and then transferred into the chamber of the counter (Figures 8.5(d) and (e)). The chamber is constructed with a plastic scintillant and the β activity of the ^{85}Kr tracer is counted by a scintillation technique (Section 5.4.2). The measured activity a (Bq) refers to the gas in the counting chamber (volume v). Hence, the effective rate of sampling of the gas from the pipeline q (l/s) into the counter is

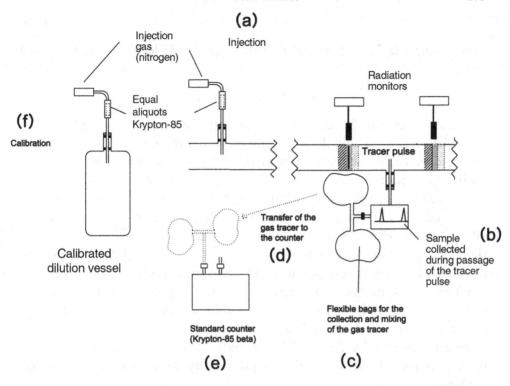

Figure 8.5.   Gas flow using tracer dilution techniques. (a) Injection of an accurately monitored aliquot of [85]Kr as an instantaneous pulse using high-pressure nitrogen. (b) Collection of a sample of the gas containing the complete isotope pulse. This is achieved by locating radiation detectors upstream and downstream of the sampling point and monitoring the [85]Kr gamma emission. (c) The sample inflates the flexible bags and is well mixed. (d), (e) The sample is transferred to the standard counter and the activity registered. (f) A second aliquot is added to a calibration vessel, mixed and transferred to the standard counter.

$$q = v/\Delta T. \tag{8.5}$$

Optimising accuracy

In principle, the gas flow rate could be calculated from Eqs. (8.4) and (8.5) if the activity ratio A/a and the counter volume v were known with sufficient accuracy. In practice, estimates of these parameters would introduce significant errors. Fortunately the uncertainties may be greatly reduced by accurately metering a second aliquot of the gas of equal activity A from the same stock supply into a calibration vessel of volume V (Figure 8.5(f)). The accurately diluted tracer is transferred from the calibration vessel to the same counting chamber to determine its activity a_c (Bq). Hence a_c/A is equal to the dilution factor v/V, i.e.

$$A/a_c = V/v. \tag{8.6}$$

Substituting the Eqs. (8.5) and (8.6) into Eq. (8.4) leads to the following expression

$$Q = (a_c/a)(V/\Delta T). \tag{8.7}$$

Gas flow rates can be determined with precision as V and ΔT can be accurately measured, and the scintillation counter is required only to measure the ratio of the specific activities of the ^{85}Kr samples (a_c/a) which are of similar magnitude.

If required, the mass flow W (kg/s) of the gas can be obtained from the expression $W = \rho Q$ where ρ is the density of the gas at ambient pressure in the dilution vessel (Figure 8.5(f)) and not in the high-pressure pipeline. This is of significant practical advantage as the densities of gases, and especially of gas mixtures, are not well known when at elevated pressures.

When this technique was employed for measurements in a natural gas pipeline it produced results within 1% of those obtained with a reference orifice plate (a standard calibration procedure), so illustrating achievable accuracies, given sufficiently careful procedures. The advantage of the tracer method is that it can be readily applied away from installed calibration facilities.

8.3.6 Flow rate measurements: total count method

The essential feature of the total count method is that flow rate is determined from the cumulative response of a strategically located detector. The detector may be immersed in the liquid, e.g. for the gauging of rivers and streams, or it may be located externally, e.g. when flow rates are monitored in pipelines. In either case the total number of counts N, corrected as required, is registered during the passage of the complete pulse. It may be shown that the flow rate Q (l/s) is given by

$$Q = AF/N \tag{8.8}$$

where A (Bq) is the total injected activity and F (cps per Bq/l) is the calibration factor relating the injected activity to the total number of counts registered during the passage of the radioactive pulse (Appendix 4, Eq. (A4.11)).

The application of the total count method to river flow measurements is described in Section 9.3.2. The calibration of fully immersed detectors involves recording their responses (cps) to known tracer concentrations (Bq/l) in a large tank (Figure 9.2(a)). When measuring flow rates in pipelines, the detector is placed external to the pipeline (Figure 9.2(c)). Calibration is effected by setting up a section of similar pipeline and measuring the response of the similarly located detector to a known activity of tracer within the pipe. Care must be taken to ensure that the geometries of the field measurement and the calibration are as similar as possible.

8.3.7 *Residence time distribution*

The efficiency of many industrial processes depends on ensuring optimum contact between the various reactants or between reactants and catalysts. If contact times are too short, the reactions could be incomplete and the yields of product reduced; if the contact times are longer than required, unwanted secondary reactions could cause problems. Examples of processes where contact times are important include:

catalytic conversions in the chemical and oil refining industries
leaching processes in the minerals extraction industry, and
the aeration of dispersed sewage in inland disposal ponds.

Radiotracer techniques have been widely used to assess the performance of industrial plants against design specifications. In doing so, attention must be paid to both the mixing processes and the residence time distributions (RTDs) of the reagents and the tracers. The concept of the RTD was introduced in Section 8.3.1 and is discussed in Section A4.7 of Appendix 4. The mixing and transport of reagents and products through an industrial plant are invariably complex. Detailed interpretation of the behaviour of the tracer can often only be made with the aid of advanced computer modelling. However, the experimental procedures can be illustrated using 'plug flow' and 'stirred flow' approximations.

8.3.8 *Residence time distribution: idealised plug flow*

Idealised plug flow (Figure 8.6(a)) assumes a completely regular, steady flow through a pipe or channel. The principles have been applied to inland sewage ponds which are designed to ensure that aeration processes can go to completion without the need for mechanical mixing and with only negligible dispersion. For plug flow, the mean residence time (Section 8.3.1), MRT is a measure of the time that components of the system remain in contact. It is given by

$$MRT = V/Q \qquad\qquad (8.9)$$

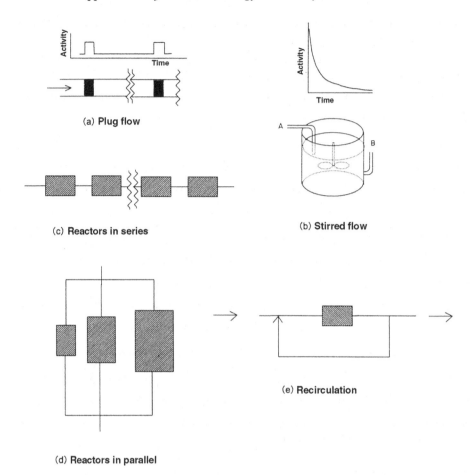

Figure 8.6. RTD under a range of flow conditions. (a) Idealised plug flow, and (b) idealised stirred flow showing the variation of activity with time following the injection of a radiotracer pulse. (c) Reactors in series. (d) Reactors in parallel. (e) Reactor with recirculation (feedback).

where V is the volume of the system and Q is the mean flow rate.

The MRT of contact of the sewage with aerated water in a sewage pond prior to release to the environment is one of the key design parameters. Plug flow is normally maintained with a series of baffles. If a system develops faults, a 'short circuit' could occur when a part of the sewage by-passes one or more of the baffle plates.

A pond with a design residence time of 24 hours was studied using the radiotracers technetium-99m and tritium (HTO). Normally the tracer is injected at the input of the pond and monitored at the output. ^{99m}Tc was chosen because (a) it is a γ ray emitter ($E_\gamma = 140$ keV) which can be continuously recorded using immersed gamma ray probes; (b) it has a conveniently short half life (6 h); and (c) it can be prepared in the field by elution from a ^{99}Mo/^{99m}Tc generator (Section 1.5.2).

As the parent ^{99}Mo has a half life of 66 h, a number of ^{99m}Tc investigations may be undertaken over a period of a week or more before the generator needs replacing (Section 1.5.1). However, the use of ^{99m}Tc has the disadvantage that it readily adsorbs onto the sewage particles and may therefore not accurately reflect the transport of the aqueous component. Tritium as water (HTO) was therefore added and samples collected periodically for laboratory assay. By monitoring the HTO/^{99m}Tc ratio, duly allowing for ^{99m}Tc decay, any significant differences in the transport of the sewage particles and the water would be apparent.

In the present example, it was found that the first evidence of tracer at the outlet appeared not 24 hours after injection as expected but after only six hours. In this case partially untreated sewage was being discharged to a river system. A detailed analysis of the tracer response made it possible to arrive at an engineered solution.

8.3.9 Residence time distribution: idealised stirred flow

Stirred flow

A schematic representation of a stirred flow reactor is shown in Figure 8.6(b). A single mixing tank operating continuously has both the liquid and, when required, a radiotracer entering at A and leaving at B (see below). Stirring must be very efficient since the mixing of the components is to occur almost instantaneously. If a radiotracer (activity A Bq) is injected as a pulse into a vessel of volume V, the initial concentration is $C_0 = A/V$ (Appendix 4, Eq. (A4.13)). In practice it is necessary to allow an adequate residence time for chemical reactions to go to completion. To effect this it is common practice to connect a number of stirred tanks in series. Mathematical expressions for the RTDs are discussed in Appendix 4 (Eqs. (A4.15) to (A4.17)).

An example from the gold extraction industry

A tracer study was commissioned to investigate the lower than expected yields in a gold extraction plant. The process was designed to extract gold with an efficiency of 92 to 96% from ore with a gold loading of a few parts per million (ppm). The plant comprises ten well stirred tanks in series. In the first five tanks gold is leached from the crushed ore with a solution of sodium cyanide (0.02 to 0.05% NaCN). The second sequence of tanks contains a slurry of activated charcoal onto which the gold precipitates at levels of 9 to 12 kg gold per metric ton of carbon. Finally, the gold is separated from the carbon using concentrated alcoholic alkaline cyanide from which it is recovered by electrolysis. The carbon is reused following its re-activation by controlled roasting.

In the present case it was surmised that the reduction in the yield was due to short circuiting of solution through one or the other of the sequences of leaching or adsorption tanks. An investigation of the RTD of the aqueous phase was therefore undertaken. The radiotracer tritium was used. Separate studies were made of the performance of the adsorption and the leaching tanks.

The adsorption tanks behaved as predicted. The tritium profiles (shown in Figure 8.7) closely reflected the mathematical expressions for the concentration profiles for stirred tanks in sequence (Appendix 4, Eq. (A4.17)). Independent estimates of the MRTs were obtained from the shapes of the concentration profiles and also from the tank volumes and the known flow rates (mean residence time = tank volume/flow rate through the tank). There was good agreement between both estimates. Each indicated mean residence times of about 90 min per tank. So far, so good.

However, the behaviour of the leaching tanks was very different. The tritium concentration profile at the outlet of each tank is depicted in Figure 8.7. Clearly the system was exhibiting anomalous behaviour. The most direct evidence for short circuiting was shown at the outlet of the second leaching tank. A relatively slow build-up of tritium was expected. In practice, high concentrations of the tracer appeared at the exit of tank 2 a few minutes after being injected at the inlet of tank 1.

A similar behaviour was observed in subsequent tanks. It appeared that a small fraction of the tracer was skimming across the surface of the leaching tanks and not mixing with the bulk flow. From a detailed analysis of the profiles it was concluded that about 16% of the flow exhibited short circuiting from tank 1 to tank 2, but the percentage decreased along the train.

The short circuiting was a likely reason for the reduction in yield of the extraction process. The problem could be solved by either introducing a sixth tank into the leaching train or, if practical, increasing the efficiency of mixing.

8.3.10 A comment on modelling complex flows

Flow patterns in chemical reactors are normally far too complex to be represented by plug flow or stirred flow approximations. In practice, there are two approaches to more realistic modelling. The first involves analysis of the

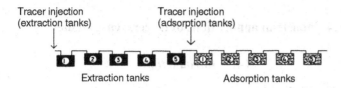

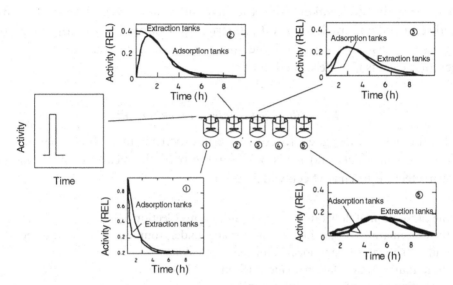

Figure 8.7.   RTD study in the gold extraction industry. A sequence of five extraction tanks and five adsorption tanks in series. An instantaneous pulse of tritium is injected at the inlet to the first extraction and to the first adsorption tank. The tritium concentrations after the first, the second, the third and the fifth tanks are shown graphically as a function of sampling times. The profiles in tank 4 may be inferred from those in 3 and 5. The extraction and adsorption tanks show significantly different behaviour.

flow patterns in terms of a number of plug flow or stirred flow elements which may either be in series or in parallel and may involve feedback circuits. The various elements are shown schematically in Figures 8.6(c), (d) and (e). The second approach involves the holistic modelling of the flow patterns using computer fluid dynamic (CFD) codes. The use of these codes presents new challenges and opportunities for tracer technology as a powerful method for experimental validation of the model output. However, detailed discussion is beyond the scope of this book.

8.4 Industrial applications of tracers: case studies

8.4.1 Introduction

The range of applications of radiotracing to industry is indicated in Table 8.3. Three examples have been selected to illustrate the diversity of investigations which may be undertaken with the help of radiotracers. The studies are fluidised catalytic cracking in oil refineries, an example of the use of radiotracers in blast furnaces, and the monitoring of industrial inventories of mercury; each is now described in turn.

8.4.2 Fluidised catalytic cracking unit

Oil refineries are designed to convert crude petroleum to fuel, heating oil, lubrication products and feedstocks for the petrochemical industry. Radioisotope techniques have been widely used for:

- the calibration of flow rate measurement systems (Section 8.3.2);
- the examination of pipelines on the surface, underground or sub-sea to detect the build-up of deposits and locate blockages or leakage;
- the scanning of distillation columns (Section 7.2.1);
- the optimisation of a range of catalytic processes.

Catalysts are used in the refining industry for: (a) catalytic cracking, i.e. the production of gasoline range hydrocarbons from the higher boiling fractions of crude oil; (b) isomerisation i.e. the conversion of low octane C_5 and C_6 n-paraffins into higher octane branched isomers; (c) reforming, i.e. the conversion of C_6+ aliphatic hydrocarbons to aromatics and branched chain hydrocarbons and thus increasing the octane number of the naptha feedstock; and d) for a range of processes in the petrochemical industry.

This section will briefly describe applications of radiotracing to fluidised catalytic cracking, a process that is central to the operation of modern petroleum gasoline refineries. In United States refineries, the amount of feed processed by fluidised catalytic cracking units (FCCUs) was, in 1991, equivalent to 35% of the total crude oil processed. A schematic representation of an FCCU is shown in Figure 8.8.

It is critical to the economics of the refinery that FCCUs operate to peak specification. Investigations are complex because there are four major process streams which come together to effect the catalytic cracking reactions: fresh feed, catalyst, steam and air. Radiotracer techniques offer a uniquely

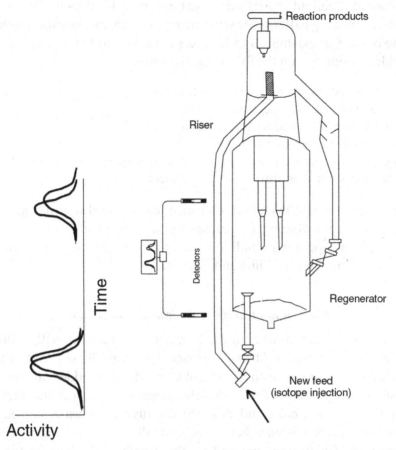

Figure 8.8.   A schematic representation of an FCCU illustrating the investigation of the dynamics of the gas phase and the catalyst in the riser.

powerful approach for investigating sub-optimal behaviour because of the following:

- Different process streams can be separately labelled and independently monitored. For instance, krypton-85 is normally used as the gas phase tracer. The catalyst may be made radioactive by irradiation in a nuclear reactor.
- Since the detectors are placed externally, the tests are performed under normal operating conditions so that interference with production schedules is minimised.
- FCCUs are too complex to enable sufficiently precise diagnostics using mathematical models alone.

In an investigation up to fifty radiation detectors are placed at strategic locations throughout the refinery plant and used to measure flow velocities, RTDs and mixing patterns for separate process streams through key elements of the plant. For example, a radiotracer injected into the riser can be used to provide information on the following processes:

feed and catalyst distribution in the riser feed zone;
catalyst and vapour velocities in the riser (slip factors);
determination of the efficiency of the riser termination device;
RTDs through the reactor and stripper;
mixing and flow characteristics in the reactor and stripper; and
cyclone efficiencies, flow distribution and residence times.

These are among the many tests which may be used to investigate a range of generic issues including: (a) reasons for reduced plant yield; (b) the impact of changes in operating conditions designed to enhance yield; and (c) the accuracy of process modelling and simulation.

8.4.3 Radiotracers in the iron and steel industry

Applications of radiotracers to the iron and steel industry will be illustrated here by an investigation of blast furnace operation. Blast furnaces are used to reduce iron ore to iron using coke to which is added a flux such as limestone or quartzite. They comprise three sections: (a) the shaft which takes the iron ore, coke and flux; (b) the tuyeres or water-cooled nozzles located around the furnace below the shaft through which air is blown into the furnace at high pressure; and (c) the hearth at the base of the furnace from which the molten iron and slag are tapped (Figure 8.9(a)). The reduction of the iron ore occurs primarily through reactions with carbon monoxide (CO). The efficiency with which the furnace works depends on many factors including the quality of the coke, the uniform mixing of the carbon monoxide gas with the iron ore (i.e. the absence of channelling) and the efficiency of the hearth drainage.

The example quoted here describes an investigation of reduced yield due, it was believed, to the development of a cold spot within the hearth furnace. Two radiotracers, ^{198}Au and ^{60}Co were injected through two air blast nozzles (tuyeres) located symmetrically with respect to the tap hole (Figure 8.9(b)). During the tapping process, samples were taken every two minutes and assayed simultaneously for the two isotopes using a γ ray spectrometer. If the furnace is behaving in accordance with specification, the response curves should overlap, reflecting similar residence time distributions in symmetrically located regions of the furnace. If this is not so, the response curves

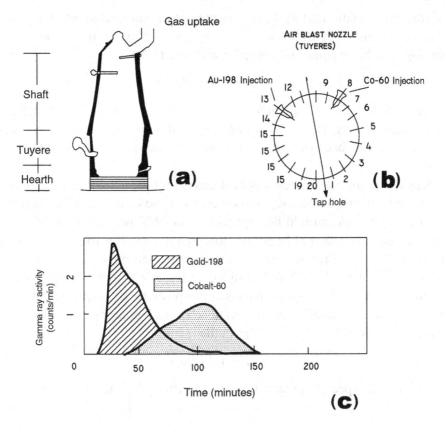

Figure 8.9.   Radioisotope study: blast furnace efficiency. (a) Schematic representation of a blast furnace showing an air blast nozzle (tuyere) and a tap hole. (b) The location of the tuyeres, the tap hole and the isotope injection points. (c) The variation with time of the activity of cobalt-60 and gold-198 in the iron sampled from the tap hole (after Easey, 1988).

differ, indicating problems most likely due to the development of cold spots within the hearth. Evidence for non-symmetrical behaviour and therefore of an operational problem is clearly shown in Figure 8.9(c).

8.4.4 Inventories

Accurate measurements of activity ratios

The use of tracer dilution techniques to measure inventories has some similarities to the determination of flow rates by the constant injection method. Both depend on the injection of a small accurately known quantity of tracer of specific activity C_0 (e.g. countrate per unit mass) into the bulk and the measurement of the much lower specific activity c following complete mixing. The basic equation (compare with Eq. (8.3)) is

$$M = mC_0/c \qquad (8.10)$$

where M and m are the masses of the total inventory and the added radioactive tracer respectively ($M \gg m$) and C_0/c is the ratio of specific activities after background and other necessary corrections have been made.

Key to the success of these applications is that the activity ratios are readily measured with high accuracy. However, the ratio of the activities before and after dilution C_0/c, should be kept within, say ± 20 per cent of unity, and the absolute activities should be neither too high nor too low to permit accurately reproducible countrates. The most common method of ensuring that the injected activities C_0 (Bq) and sampled activities c (Bq) are comparable is to accurately dilute the high activity sample prior to the measurement with a non-radioactive material of the same chemical composition. Assuming a dilution factor of k_D, Eq. (8.10) takes the form:

$$M = mk_D\, c_0/c \qquad (8.11)$$

where c_0 is the specific activity of the tracer after dilution, i.e. $C_0 = k_D\, c_0$.

Mercury inventories

Tracer dilution methods are widely used for the accurate measurement of mercury inventories in industrial plants designed for the production of chlorine and sodium hydroxide by the electrolysis of sodium chloride solutions. Although mercury electrodes are now no longer installed, many older plants are still in operation and must be carefully monitored because of the potential environmental impact of mercury release.

The radioactive mercury is produced in a research reactor by the irradiation of stable mercury. Stable mercury is composed of seven isotopes, but in practice only ^{196}Hg and ^{202}Hg form significant irradiation products. Which of the two is produced with the relatively greater intensity depends on the irradiation time. The resulting radioisotopes are ^{197}Hg and ^{203}Hg (Table 8.1) which have half lives of 64.1 h and 46.6 d respectively.

The method is simple in principle. The radioactive mercury, either ^{197}Hg or ^{203}Hg, of accurately known mass m and specific activity C_0 is added to the large reservoir, the mass M of which is to be determined. After complete mixing, a sample is taken and the specific activity c measured. A second aliquot from the activated solution is diluted by a factor of k_D yielding a specific activity of c_0. The inventory is then calculated from Eq. (8.11).

Which radionuclide is used for the investigation is determined by the time required to achieve a homogeneous mixing of the added radioisotope with the mercury in the reservoir. If this time can be expected to be no longer than a day or two, irradiation conditions are chosen to maximise the yield of the shorter lived ^{197}Hg. This approach is cheaper and leads to less residual radioactivity. If complete mixing requires several days, the ^{197}Hg is allowed to decay and the ^{203}Hg is used as the tracer. The standards, samples and blanks are counted under reproducible conditions using a γ ray spectrometer. With careful work, measurement accuracy better than 1% can be achieved routinely.

8.5 Conclusions

Radiotracer techniques have been extensively used throughout industry since the mid-1950s to: (a) obtain an improved understanding of industrial processes; (b) diagnose the reasons for industrial plant operating below specification; and (c) obtain accurate measurements of flow rates, inventories and mixing efficiencies.

A number of factors have influenced the evolution of radiotracer technologies over the past four decades. First and foremost has been the enormous developments in microprocessor technology, data processing and visualisation. Some classes of modern investigations require the rapid accumulation and modelling of data from fifty or more detectors.

However, radioisotope applications are highly regulated which complicates their employment and means that alternative diagnostic technologies using unregulated tracers are often preferred even when they have technical limitations. Never-the-less, radiotracers remain in use for large and complex tasks for which there are no practical alternatives and also for many others.

Developments of radioisotope technologies are not standing still. With the passage of time, users are demanding more sophisticated data visualisation. Techniques developed for the health sciences are being applied to industry. For example, it is expected that computerised tomographic imaging using γ rays from radiotracers will eventually become a standard requirement in many investigations.

Regulatory provisions for work with ionising radiations are designed to ensure that doses to workers not only fully comply with all requirements, but are, in addition, as low as reasonably achievable (the ALARA principle). The challenge is therefore to maximise the amount of information which can be obtained with the minimum use of radioactivity. Tracers are now seldom used in isolation, but in conjunction with a range of investigation techniques. They are increasingly seen as an essential aid to achieving the fundamental aim of extending and validating numerical models of complex industrial systems.

Chapter 9

Radionuclides to protect the environment

9.1 Introduction

The last chapter of our book will describe the use of radioisotopes for assessing the impact of human activities on the environment. The range of opportunities is very extensive as seen by the list of applications in Table 9.1 which is by no means complete. Emphasis will be placed on assessing such impacts using two methods, described here as the mathematical modelling approach and the archival approach.

1   *The modelling approach.* Due to recent vast increases in the power of computers, applications of mathematical modelling techniques to environmental science are now very widespread and underpin much of our detailed understanding of processes in the oceans, the atmosphere and coastal and terrestrial ecosystems. Since increasing use is being made of numerical models for scientific prediction and environmental management, there is a growing need for independent verification of the model predictions. Tracer techniques are a powerful tool for model validation, as will be demonstrated in several examples.

2   *The archival approach.* This approach to the study of environmental processes involves using evidence from the past to understand the present and to assess the future. The approach is implemented in two stages. The first stage involves the systematic dating of material which has accumulated layer by layer often over long periods of time. Cores extracted from sediments, ice sheets, old growth trees or massive corals are commonly used. From knowledge of the dates of the samples and their locations in the core, it is possible to calculate the average rate of growth or accumulation. This leads to the second stage of the archival approach, which involves the measurement and interpretation of indicators of past environmental change which are retained in the samples.

The age determination of samples using naturally occurring or environmental isotopes involves three steps: (a) laboratory measurement of the activities of the isotopes in carefully selected and prepared samples; (b) an assessment of the

Table 9.1. *Applications of tracer technology to the environment (see Tables 8.1 and 8.2 for decay data).*

No	Environmental Sector	Investigation	Environmental Impact	Radioisotope technique	Comments
1.1	*Environmental science and engineering* Contaminant dispersion	Evaluation of ocean sewage outfalls	Potential impact on public health, on the recreational uses of beaches, on fisheries, etc.	Artificial radiotracers (tritium (HTO), ^{198}Au or ^{99m}Tc) used to study the dynamics of the processes, and to validate numerical models	See Sections 9.3.3 and 9.3.4
1.2		Dynamics of inland sewage ponds	Potential for impacting on the local river systems or groundwater	Tritium and ^{99m}Tc tracers used to measure residence time distributions and identify 'short circuiting'	See Section 8.3.8 Essential that the residence times of the sewage ponds are sufficient to ensure complete aerobic oxidation of pathogens
1.3		Ecological impact of contaminant dispersion	The degradation of ecosystems must be avoided if sustainable development is to be achieved	A wide range of radionuclides are available e.g. ^{65}Zn, ^{64}Cu and ^{54}Mn used to study uptake of heavy metals, ^{32}P for phosphates, etc.	
1.4		Discharges of air pollutants from industrial stacks	Degradation of air quality in the vicinity of the stack	Validation of stack dispersion models	Chemical tracers detectable with high sensitivity are normally used

2.1	*Coastal and hydraulic engineering* Applications to dredging	Optimisation of location of dumping sites for dredge spoil	If not optimally located, the dredge spoil might for example migrate into a shipping channel, or contaminate the coastline	Dredge spoil labelled with e.g. ^{198}Au, ^{46}Sc and tracked	
2.2	Harbour and port development	Measurements of the migration of sand and sediment to e.g. optimise the alignment of dredged shipping channels or sea walls	Failure to optimise alignment will give rise to a need for ongoing maintenance, dredging and other negative impacts	Sand and sediment tracing (^{198}Au, ^{192}Ir, etc.) over periods from 1 week to 1 year used to validate transport codes	Section 9.3.5
2.3	Hydraulics	Gauging of rivers		Tracer dilution methods for calibration or where gauging stations are not available	Section 9.3.2
3	*Groundwater*	Efficiency of recharge to groundwater basins; the dynamics of groundwater flow	Over-exploitation or pollution of a groundwater basin can have impacts which persist for tens to hundreds of years or longer	Cosmogenic isotopes tritium, ^{14}C used to age samples and hence define recharge areas and migration rates; stable isotopes D/H and ^{18}O/^{16}O to study groundwater source and evidence for mixing.	Isotopic methods are of greatest value where extensive hydraulic data are unavailable

Table 9.1. (*contd*)

No	Environmental Sector	Investigation	Environmental Impact	Radioisotope technique	Comments
4.1	*Sedimentation and erosion* Sedimentation	Rate of accumulation of sediments in reservoirs and estuaries	Loss of capacity of reservoirs Impact on harbour development and on the maintenance of shipping channels	Sediments are cored; the cores sectioned with depth and the sections dated	The dating technique chosen depends on the time scale: ^{137}Cs up to decades; ^{210}Pb to 100 y; ^{14}C to 20 000 y; ^{230}Th/^{234}U to 250 000 y
4.2	Erosion	Rate of erosion and accumulation of soils in catchments	Impact of land use change on soil erosion and productivity	^{137}Cs profiles over catchments allow quantitative estimates of soil erosion and accumulation	Such data can be used to calibrate empirical equations designed to predict erosion rates over geographical regions
5.1	*Oceanography*	Transport of solutes and particulates in the water column		Profiles of 'bomb' isotopes, analysis of U, Th and Ra isotopes in the water column; ^{230}Th /^{234}U in sea bed	Applications widespread, e.g. impact of rivers on coastal zone, transport of nutrients, development of coral reefs, etc.
5.2	Ocean currents	Validation of ocean current models	Global ocean currents directly impact climate	^{14}C used to measure the residence times water bodies	Residence time is the time elapsed since the ocean water sample was at the surface
6.1	*Global climate change* Atmospheric transport	Validation of atmospheric transport models	Improved prediction of postulated 'greenhouse' effects	Radon used to establish whether air mass passed over land	Improved prediction of postulated 'Green-house' effects

6.2	Glaciology	Ice cores dated and provide an archival record of global temperatures and the levels of 'green-house' gases		Stable isotopes D/H $^{18}O/^{16}O$ (temperature and dating) and ^{14}C techniques correlated with the analyses of gases from ice cores	Accelerator Mass Spectrometry (AMS) essential as small samples normally only available Information on past climate also from sediment cores
7.1	*Nuclear waste disposal* Natural analogues of waste repositories	Validation of performance assessment models over long time scales	Possible impact of the leakage of radionuclides on future generations	Uranium ore bodies as analogues of spent reactor fuel in waste repositories	Uranium series nuclides, e.g. $^{234}U/^{238}U$, $^{230}Th/^{234}U$, $^{226}Ra/^{230}Th$ isotope ratios in host rock and groundwater near a deposit used to validate aspects of radionuclide transport models

initial values of the activities at the time when the samples were formed; (c) calculation of the age of the sample from the mathematical relationship linking the initial and the measured values.

Dating techniques are most reliable when, as in the case of carbon-14, the age of a specimen is determined by the decay of a single isotope. If, on the other hand, the isotope is a member of a decay chain, the build-up and decay of the parent (Section 1.6.2) may also need to be taken into account. The interpretation is usually more complex and so are the measurements. A large number of techniques for the dating of geological samples have been developed (Geyh and Sleicher, 1990). Although geochronology is outside the scope of this book, Table 9.2 contains a list of the commonly used dating methods.

The second stage of the archival approach involves measurement of indicators of environmental change in the dated samples. For instance, the systematic monitoring of levels of heavy metal pollutants in dated sediment or coral cores may provide important information on the history of effluent release in the local region. Again, systematic variations in the $^{18}O/^{16}O$ and D/H ratios in ice cores reflect changes in global temperature and therefore in global climate. Several applications are discussed in more detail in Section 9.3.

9.2 The investigation of environmental systems

9.2.1 Numerical modelling

Environmental investigations have traditionally involved a combination of data gathering and scientific interpretation. In more recent times, computing has developed to the extent that a third element, numerical modelling, is of comparable significance. Its impact has been greatly enhanced by parallel developments in the visualisation of data. Reference need only be made to modern graphical representations of results obtained in the laboratory and in the field, as well as the visualisation of satellite images and the output of numerical codes. Spectacular examples of the latter include the visualisation of global climate change models and ocean circulation models.

The rapid growth in the use of numerical codes has led to a shift in the role of tracer techniques. As with industrial applications, tracers are now used less for the investigation of particular processes, and more for model validation. To illustrate this point, reference will be made to studies of contaminant dispersion in Section 9.3.3.

Many environmental systems cannot be modelled satisfactorily, either because they are inherently too complex or because too little is known of the underlying physical processes. In such cases statistical or correlation techniques are often helpful in establishing the relative importance of different

Table 9.2. *Environmental radioisotopes.*

Isotope	Source	Half life	Analysis	Application	Comments
H-3 (tritium)	Atmospheric testing cosmogenic	12.3 y	β counting (liquid scintillation, Section 5.4.2)	Groundwater recharge. Oceanographic mixing	Environmental tritium dominated by atmospheric testing source; hence tritium in water implies that a component of the water from post-nuclear i.e. post-1950 precipitation. Environmental tritium is used to identify groundwater recharge (Section 9.4.3) areas and to study oceanographic surface mixing processes (Section 9.4.4)
Be-7	Cosmogenic	53 d	γ spectrometry at 480 keV (Section 6.4)	Sediment accumulation and redistribution over the previous half year	The technique for Be-7 counting is similar to that for Cs-137 counting (Section 9.4.2). The samples are dried, weighed, placed in a Narinelli beaker over a high-resolution detector and the 480 keV gamma peak measured (see Section 9.4.1) The presence of Be-7 in a sediment core indicate that the material has been at the surface over the past six months. Be-7 is correlated with Cs-137 in sediment cores which provides information on the rate of accumulation of sediment in post nuclear times i.e. over the past 40 years
Be-10	Cosmogenic	1.6×10^6 y	AMS		Be-10 is used to study sediment accumulation and redistribution over the past few million years. Where possible, it is used in association with Al-26

Table 9.2. (*cont.*)

Isotope	Source	Half life	Analysis	Application	Comments
C-14	Atmospheric testing	5730 y	AMS, β counting	Post-nuclear processes (hydrology, oceanography)	In post-nuclear times, atmospheric nuclear testing is the major source of C-14. It therefore exhibits a typical 'bomb' pulse which has been used, for example: ■ to study the uptake of CO_2 by the oceans; to investigate mixing processes in the upper layers of the oceans in post-nuclear times; ■ to study the incorporation of atmospheric CO_2 into antarctic ice firn in order to refine the C-14 age dating of ice cores; and ■ to provide independent evidence of the source of carbon (modern vegetation or mineral) in commercial products should it be contested
	Cosmogenic			Pre-history; evolution of coastal and other ecological systems in recent geological time; global climate change studies	C-14 is used for dating carbon containing materials up to 50 000 y. The technique has been widely used for: ■ dating artefacts, bones and charcoal and thereby making a major contribution to an understanding of pre-history and the evolution of ecological systems; ■ studying groundwater flow patterns; ■ investigating coastal processes through, for instance, the dating of marine corals, of shell grit in dunes and of sediments in estuaries and lakes;

Isotope	Origin	Half-life	Method	Application	
					■ better understanding of global oceanographic circulation patterns; and ■ dating of tree rings, ice cores and corals as a contribution to global climate change studies
Al-26	Cosmogenic	740 000 y	AMS	Erosion and sedimentology	Al-26 together with Be-10 is used to study the rate of accumulation of sediments and of the erosion of exposed rocks over millions of years
Cl-36	Atmospheric testing Cosmogenic	301 000 y	AMS	Salinity, groundwater quality	The Cl-36 bomb pulse is used to chloride migration i.e. salinity processes in the unsaturated zone and in modern groundwater
				Dating of old groundwater	Cl-36 is used to date groundwater up to one million years old. More generally it is used to study aspects of the chloride cycle, i.e. the evolution of groundwater quality
Cs-137	Atmospheric testing	30.1 y	γ spectrometry	Sedimentation, soil erosion	Cs-137 is used to measure the rate of sedimentation and erosion in post-nuclear times. Please refer to Section 9.4.2 and to Table 9.1
Th-232 U-238 U-235	Primordial Primordial Primordial	1.4×10^{10} y 4.47×10^{9} y 7.0×10^{8} y	α spectrometry, Mass spectrometry	Dating geological samples	Th-232, U-238 and U-235 are the progenitors of Pb-208, Pb-206 and Pb-207 respectively. They are used to date old geological material of age up to 10^{9} years or more. Ion microprobe methods are available to date individual inclusions within natural materials and have been widely applied to the dating of zircons

Table 9.2. (*cont.*)

Isotope	Source	Half life	Analysis	Application	Comments
U-234	U series	245 500 y	γ–X spectrometry, Thermal Ionisation Mass Spectrometry	Uranium migration; age sedimentary U deposits	$^{234}U/^{238}U$ ratios are measured in geochemical samples and groundwater. Together with other data they are used to study the dissolution, migration and deposition of uranium. Applications include the age and stability of uranium deposits
Th-230	U series	75 380 y		Dating sediments and corals	$^{230}Th/^{234}U$ activity ratios have been widely used for the dating of the accumulation of sediments in estuaries and to the dating of massive corals (e.g. porites)
Ra-226	U series	1601 y	γ–X spectrometry		Radium-226 is the parent of radon-222
Rn-222	U series	3.83 d			The level of radon in houses and in drinking waters is extensively monitored since it could be a significant source of the annual radiation dose received by the general public (Table 2.5)

Radon is emitted from the soil and is a natural tracer for air masses which have passed over land within the preceding few days. Air masses which have not 'seen' land for a few days are much lower in radon. Their levels are measured in a number of reference stations and contribute to the validation of atmospheric circulation models and hence to climate change studies |

| Pb-210 | U series | 22.2 y | Sediment dating over the past 100 y | The ^{210}Pb method has been extensively applied to the dating of sediments in lakes and estuaries over the past 100 years. The sediment is cored and individual sections assayed for ^{210}Pb. There are two components of ^{210}Pb, the unsupported and the supported. The unsupported lead, which is used in measuring the accumulation rates, is adsorbed on the surface from the decay of the ^{222}Rn dissolved in the associated water. The supported ^{210}Pb is derived from the decay of the ^{238}U within the sediment minerals. Experimentally, the supported lead is calculated from the measured uranium levels, and is subtracted from the total to obtain the 'unsupported' component |

parameters. For instance the rate of erosion depends on a wide range of factors including the intensity and distribution of rainfall patterns, the properties of soils and the nature and coverage of vegetation. The most widely used equation for estimating soil erosion is the Universal Soil Loss Equation (USLE), which was developed for soils east of the Rocky Mountains in the United States (Wischmeier and Smith, 1965). As discussed in Section 9.4.2 this equation has often been used in settings where it does not strictly apply. Some form of validation is essential. In this case, the distribution of environmental isotopes such as the fall-out product ^{137}Cs may be used to provide additional independent measurements of erosion rates at different points within a catchment. Such information can be used to refine the correlation equations and enhance their predictive value.

9.2.2 Applications of radioisotopes

Both man-made and naturally occurring radioisotopes are used extensively in environmental science. Many applications of the former have evolved out of analogous studies of industrial processes discussed in Chapter 8. For instance, the use of tracer dilution techniques for flow rate measurements which were developed for pipeline studies (Section 8.3.2) have been widely applied to the gauging of rivers and streams (Section 9.3.2).

Naturally occurring radionuclides, also known as environmental radionuclides, are classified into three sub-groups according to their source – cosmogenic radioisotopes, 'fall-out' products and primordial isotopes.

1. *Cosmogenic radioisotopes* notably tritium, beryllium-10, carbon-14 and chlorine-36 are generated principally in the upper troposphere and lower stratosphere by the impact of cosmic rays on the components of the residual atmosphere. They subsequently diffuse to the surface of the land and the oceans either with rainfall or associated with particulates.
2. *Fall-out products* result from past nuclear testing in the atmosphere. They include a range of fission products, such as caesium-137 as well as tritium, carbon-14 and chlorine-36 which are also cosmogenic isotopes. Their yields reached a maximum in the mid-1960s prior to the Atmospheric Test Ban Treaty and then decreased. This pattern is often observable in environmental systems and is referred to as the 'bomb pulse'.
3. *Primordial isotopes* such as uranium-238, thorium-232 and their daughters (Figure 1.5) have been present since the formation of the earth. The stable isotopes deuterium, carbon-13 and oxygen-18 could also be included within this group.

Each class of radionuclides has particular features which may be exploited in the design of an environmental investigation. For instance, studies involving artificial radioisotopes can be precisely tailored to particular processes. A number of examples are presented in Table 9.1. On the other hand environmental isotopes are widely distributed tracers and can be used to study the cumulative effects of processes over a wide geographic area (Table 9.2). Accessible time scales vary widely depending on the half life of the isotope and the sensitivity with which it can be measured. Of course, fall-out products are the exception as they only entered the environment in post-nuclear times.

9.3 Environmental applications of radioisotopes

9.3.1 Introduction

Applications of reactor-produced radioisotopes to industry were discussed in Chapter 8 and listed in Table 8.3. As noted in Section 8.1.3, some of the earliest environmental applications of tracers were to the measurements of river flow, sewage dispersion and sand and sediment migration.

Radioisotope investigations are becoming increasingly sophisticated, requiring precise quantitative data. Procedures for accurate activity measurements in laboratories were discussed in Sections 6.3 and 6.4. As a rule, the calibration of field detectors also has to be done in the laboratory when it is often difficult to reproduce the counting geometry used in the field. The calibration of detectors for river flow and sand transport measurements is discussed in Sections 9.3.2 and 9.3.5 respectively.

9.3.2 River flow measurements

The total sample method using tritium

The application of tracer dilution techniques to the measurement of flow rates in industry was discussed in Section 8.3. Reference was made to the total sample and the total count techniques. Both methods have been applied extensively to the gauging of rivers and streams. The principle is illustrated in Figure 9.1(a). The isotope is injected as an instantaneous pulse at X and, after complete mixing has been achieved, the plume is monitored at a convenient sampling point Y.

Chemically, tritium as HTO is the 'perfect' tracer for water. Furthermore, it can be easily transported in a sealed glass vial, and requires a minimum of shielding. Care must be taken to ensure that investigators do not inhale any tritiated water vapour when releasing the tracer into the river, but such a precaution is readily achieved. Experimental procedures are designed to be as

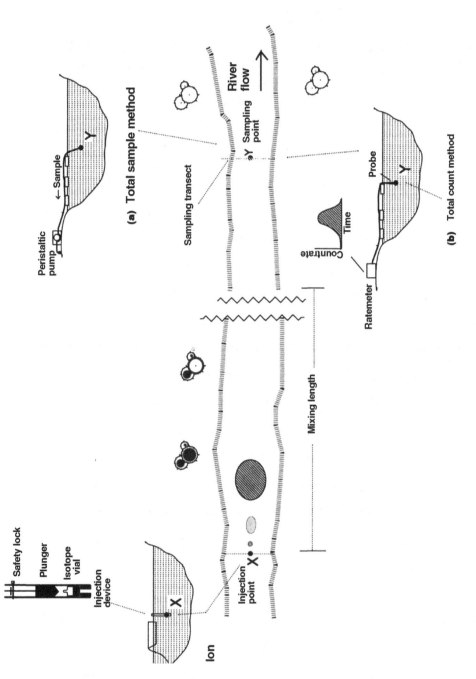

Figure 9.1. The gauging of rivers using dilution methods following the instantaneous release of the radiotracer with the injection device (bottle breaker). The cross sections at the injection and sampling transects X and Y are shown. (a) Total sample method: a sample is collected at an accurately known rate and the activity measured. (b) Total count method: the detector is immersed in the river and the total counts recorded during the passage of the plume.

simple as possible. For instance, injection may be effected by transferring the vial to a bottle-breaking device using remote handling equipment, immersing the device in the middle of the river below the surface and breaking the vial by releasing a heavy weight.

The river is sampled downstream at a constant rate q (l/s) (Figure 9.1(a)). Sampling is commenced before the arrival of the radioactive pulse and completed only after the pulse has passed. Assuming complete mixing has been achieved between injection and sampling, the total activity collected during sampling a (Bq) will be independent of the specific location of the sampling point. If the activity of the injected tritium is A (Bq), then the flow rate of the river Q (l/s) is given by Eq. (8.4), namely

$$Q = qA/a \tag{9.1}$$

The total count method

The principal disadvantage of the tritium method is that the diluted samples must be returned to the laboratory for activity measurements. A gamma emitting radiotracer is therefore often preferred and the *total count* method applied. The isotopes ^{82}Br or ^{99m}Tc, eluted from a molybdenum/technetium generator, are often used. Here, the data are collected in real time using field NaI(Tl) gamma detectors (Figure 9.1(b)) immersed in the stream. If N is the total number of counts accumulated by the counter during the passage of the complete pulse (corrected for background and decay), the flow rate Q (l/s) is given by Eq. (8.8), namely

$$Q = AF/N \tag{9.2}$$

where A (Bq) is the activity of the injected radiotracer and F (cps per Bq/l) is the calibration factor for the detector.

The accuracy of river flow measurements depends on the accuracy of the calibration of the detection equipment, a point referred to in Section 8.2.4 when reference was made to the problem of reproducing in the laboratory the counting geometry found in the field. In most estuarine and offshore studies, the detector probes are completely immersed in the water, a geometry which has been described as quasi-infinite. These conditions are replicated in the laboratory by positioning the detector in the centre of a large cylindrical tank and recording the countrate following additions of known activities of the isotope of interest. The tank must be large enough to avoid wall effects. The dimensions depend on the penetrating power of the γ rays. A cylindrical tank of 1 m radius and 2 m height is normally adequate to approximate infinite geometry since the half thickness in water of a 1 MeV γ ray is only about 10 cm (Figure 9.2(a)).

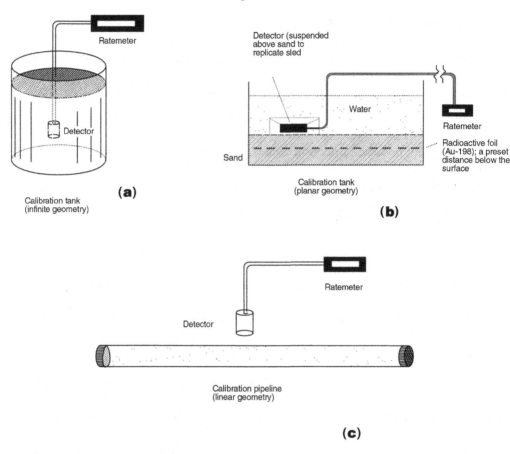

Figure 9.2.   Field calibration geometries. Three limiting cases. (a) Quasi-infinite geometry. The isotope of known specific activity is dispersed in a tank large enough to avoid wall effects. The field detector is positioned in the centre. Multiple readings are made and the calibration factor F reported as (cps per Bq/l). (b) Quasi-planar geometry. Small calibrated sources (e.g. ^{198}Au foils) are set in a uniform pattern, a fixed distance below the surface of the sand. The detector is set in place and the countrate noted. The calibration factor F (cps per Bq/m^2) is measured at a number of different depths. (c) Quasi-linear geometry. A calibration pipeline, as similar as possible to the test line, is filled with a solution of known specific activity. The detector is located as in the test and the calibration factor F measured in cps per Bq/l.

As with the total sample method, it is assumed that the conditions for complete mixing have been met (Section 8.3.1). Complete mixing is achieved when the total count during the passage of the radioactive plume is independent of the location of the detector in the river. Clearly, it is seldom possible to confirm this experimentally in a field investigation. The concept of the mixing length has therefore been introduced. If the distance from the point of injection to the point of measurement exceeds this value, complete

Table 9.3. *The mixing lengths in rivers.* [a]

Injection point	Formula	Legend	Comments
Centre	$L = 50\,Q^{1/3}$	L the mixing length; Q the volume flow	Hull formulae, reported by J. Guizerix and T. Florkowski in IAEA (1983, Ch. 5)
Bank	$L = 200\,Q^{1/3}$		
Centre	$L^{b} = 0.1\,u_x\,w/D_{\mathrm{T}}$	L the mixing length; u_x the mean velocity at measurement cross-section	After Fisher *et al.* (1979)[c]
Bank	$L^{b} = 0.4\,u_x\,w/D_{\mathrm{T}}$	w the width of the river; D_{T} the transverse disposition coefficient	

Notes:

[a] Any formula can only provide an estimate of the distance required for mixing. In situ tests are required for a firm determination of L.

[b] A degree of mixing of 95% is assumed. The degree of mixing is the extent to which mixing has been achieved in a cross section downstream from the injection of the tracer.

[c] Please also note the Draft International Standard: ISO/DIS 555-3 *Liquid Flow Methods in Open Channel – Dilution Methods for the Measurement of Steady Flow* Part 3: *Radioactive Tracers* (Published/Created: Geneva: ISO, 1991)

mixing may be assumed. A number of empirical expressions for the mixing length have been developed, some of which are listed in Table 9.3.

For further information on the gauging of rivers, readers are referred to IAEA (1983).

9.3.3 Studies of the dispersion of contaminants

Competition for environmental resources

The pollution of river systems and the coastal zone by industrial effluents and sewage is of major concern. The needs of city populations and industries to dispose of effluents must be balanced against public health issues as well as the desires of recreational users and the tourist industry for pollution-free environments. Management of these competing demands requires a robust method for predicting the impact of contaminant releases on the local ecosystems. The purpose of this section is to demonstrate that radiotracer techniques can offer powerful tools for validating predictive mathematical models, so greatly increasing the confidence with which these models can be used. Readers are referred to Lewis (1997) for further information on dispersion in estuarine and coastal waters.

Dispersion of contaminants

The transport of contaminants in rivers and other receiving waters may be studied by injecting a pulse of a radiotracer mimicking the pollutant and observing its subsequent behaviour. Close to the point of injection, the radioactive plume is carried by advective flow and it also disperses in three dimensions. The dispersion of solute (or tracer) is caused by the turbulent structure of the fluid.

Neglecting the effects of boundaries, the concentration profile of a tracer following a pulse injection resembles a Gaussian distribution (Figure 9.3(a)) – a useful fact since these distributions are well known from numerous applications.

The figure shows that there are two components to the transport of the tracer, namely advective transport and dispersion. Advective transport of a contaminant is that due to the bulk movement of the water. In this discussion, the bulk flow is in the longitudinal L (or x) direction. The advective flux ϕ_{adv, u_x} is the product of the average flow rate of the water, u_x (l/s) through a cross section, and the average concentration of the contaminant C, i.e.

$$\phi_{adv,L} = u_x\, C. \qquad (9.3)$$

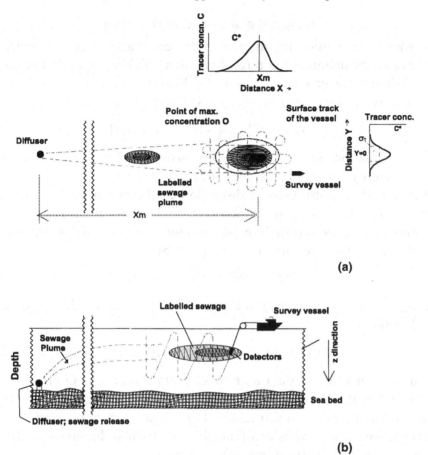

Figure 9.3.   The dispersion of tracer injected as a pulse. (a) The longitudinal (x) and transverse (y) dispersion of the pulse. The concentration profile in the transverse direction approaches a Gaussian distribution in the far field. (b) The dispersion of the plume in the vertical (z) direction. The scheme for tracking the plume in the three dimensions is shown.

   In addition, contaminants are dispersed and diluted within the receiving waters, principally as a consequence of the turbulent structure of the water. These processes are important in determining the fate and behaviour of the contaminants. The efficiency of dispersion is often quantified in terms of dispersion coefficients which are defined in terms of Eq. (9.4). Some of the basic theory is introduced in Appendix 5. It suffices to say that tracer techniques offer the most direct method of estimating dispersion coefficients in the field.

An analytical treatment of dispersion

Following the instantaneous injection of a tracer of activity A_o, concentration profiles may be monitored in three dimensions. Taking as a reference point, the maximum tracer concentration (O, Figure 9.3(a)), the profile in the transverse (or y) direction may be expressed in the form (Eq. (A5.12))

$$C(y,t) = C_y^* \exp[-(y-y_m)^2/4D_T t] \qquad (9.4)$$

where y is the location on the transverse axis from the point of maximum concentration y_m; $C(y,t)$ is the concentration of the tracer at the location y and time t; D_T is the transverse dispersion coefficient; and C_y^* is the tracer activity at the maximum concentration (O) on the transverse (y) axis.

Comparing Eq. (9.4) with the expression for the Gaussian distribution (Eq. (6.10)) it is seen that the variance (σ^2) is given by

$$\sigma^2 = 2 D_T t. \qquad (9.5)$$

The expression corresponding to Eq. (9.4) for dispersion in the longitudinal or x direction is

$$C(x,t) = C_x^* \exp[-(x-x_m)^2/4D_L t]. \qquad (9.6)$$

The maximum value of the tracer concentration occurs at O or x_m (where $x_m = u_L t$). Equation (9.6) is only an approximation, as, in practice, the tracer plumes normally exhibit a substantial 'tail' as shown in Figure 9.3(a). When the advective velocity u_L is constant, the longitudinal dispersion coefficient D_L may be replaced by the dispersivity a_L where

$$a_L = D_L/u_L. \qquad (9.7)$$

The Gaussian approximation is widely used in studies of the dispersion of contaminants in waterways. Following the release of the tracer, the radioactivity levels are tracked in a systematic way and the shape of the dispersed plume is reconstructed. The dispersion coefficients or mixing functions are calculated from Eqs. (9.4) and (9.6). Details of the calculations are presented in Appendix 5. A case study on sewage dispersion follows in Section 9.3.4.

Elaborate fluid dynamic codes have been developed to describe transport processes in rivers, estuaries and the coastal zone. However, details are beyond the scope of this text.

9.3.4 A case study: sewage dispersion

The monitoring of the dispersion of sewage and other contaminants into waterways is an application of radiotracers of demonstrated usefulness and will here serve as a case study. Many major cities around the world are located on the coastal zone, and their sewage is frequently disposed of through deep ocean outfalls following pre-treatment. The effluent is pumped a substantial distance offshore and released at intervals through a series of diffuser heads (Figure 9.4(a)). The aim of the engineering works is to minimise any impact of the sewage release on sensitive regions such as bathing beaches or the spawning grounds of fishes because of the potential impact on public health.

The evaluation of the performance of such outfalls involves the modelling of sewage effluent from close to the point of release to far into the deep ocean. Model validation is commonly effected by injecting radioactive tracers into the sewage effluent, and monitoring the dispersion of the radioactive plume for many kilometres from the underwater release point. The radiotracers are commonly gold-198 and tritiated water HTO. The period of injection usually varies from one to 12 hours.

The activity of the γ ray emitting gold-198 (Table 8.1) is measured with submerged detectors. The data are combined with those from depth meters and positioning equipment to allow the development of a three dimensional visualisation of the dispersed sewage. The tritium levels serve for an accurate measurement of the dilution factors. Water samples are taken and returned to the laboratory for measurement, together with records of the position, depth and gold-198 countrates.

The radiotracers permit detailed studies of the behaviour of the contaminants. Close to the outfall, the performance of individual diffusers can be readily distinguished (Figure 9.4(b)). The dilution factors may be measured and compared directly with specifications. Occasionally, partial or complete blockages of the diffusers are observed. The dispersion of the plume may be traced systematically into the so-called far field.

An example is presented (Figure 9.4(c)) of a study of the dispersion of sewage from the Malabar outfall, which serves part of Sydney and discharges into the ocean at 80 m depth some 4 km offshore. Plumes have been tracked in the ocean for distances up to 35 km from the point of release.

Treated sewage is a complex mixture of dissolved aqueous contaminants, suspended particulates and dispersed greases. Using radionuclides, these different components of the sewage plume can be tagged independently and studied separately. For example, the aqueous components of the sewage are labelled with tritiated water; the fine particulates with gold-198 and the organic grease component of the sewage with tritiated organic compounds. In more elaborate investigations, the levels of bacteria introduced into the water with the sewage are measured and the results corrected for dilution in

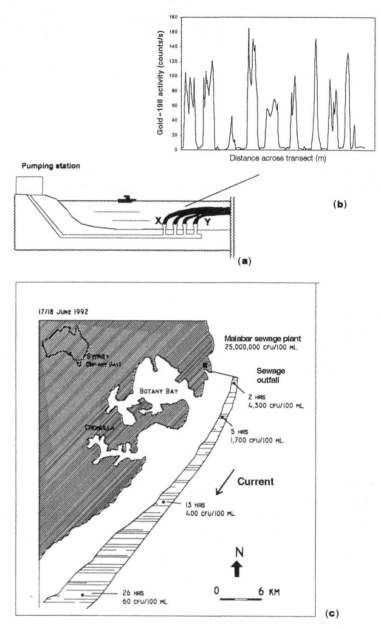

Figure 9.4.   (a) Sketch of an outfall with labelled sewage dispersing from multiple diffusers. The merging of the plume is shown. (b) The response of the radiation detector tracking close to the diffusers. The technique for tracking the dispersing plume is shown in Figure 9.3(b). (c) A radioisotope study of the dispersion of sewage from a deep ocean outfall near Sydney, Australia. Observations were made for over 26 hours. The levels of bacteria in the sea water are shown as the empirical measure colony forming units (CFU).

the ocean using the tritium measurements. It is then possible to determine a true 'die off' rate for the bacteria under the conditions found in the ocean. The results of such an investigation are shown in Figure 9.4(c) (Pritchard *et al.*, 1993).

9.3.5 *Applications of tracer techniques to sediment and sand tracing*

Measurements of migration rates

The assessment of major proposals for port and harbour development frequently calls for a sound knowledge of the offshore migration of sand and sediment. Measurements of migration rates could extend from a few days to over a year depending on the processes of interest. The radionuclides are chosen to suit the duration of the investigation. For example, ^{198}Au ($T_{1/2} = 2.7$ d) has a useful life of about two weeks, ^{51}Cr ($T_{1/2} = 27.7$ d) about four months, ^{192}Ir ($T_{1/2} = 73.8$ d) about eight months, ^{46}Sc ($T_{1/2} = 83.8$ d) about nine months and ^{110m}Ag ($T_{1/2} = 250$ d) over one year.

To label the sand or sediment, the radiotracer is either adsorbed onto the surface of the sand, or incorporated into a glass and ground and sieved to match the particle size distribution of the material of interest. The radio-labelled material is released using remote handling techniques (Figure 9.5(a)) and is monitored on the sea bed using calibrated detectors attached to a sled or other device constructed to ensure reproducible source–detector geometry (Figure 9.5(b)). It is then possible to make quantitative estimates of the rate of bed load transport of the sediments. The calibration of the detectors is discussed below.

The current status of the technology is illustrated in a long-term investigation of the migration of sand at a depth of 68 m (Figure 9.5(d)). The aim of the study was to examine the effect of storms on the movement of sand at depth. This was effected by labelling glass beads with the same particle size distribution as the sand with iridium-192 (half life 74 days). The tracer was transported to the site in three separate vials and released about 2 m from the bottom using a detonation technique. With modern position fixing techniques, changes in the activity profile of the labelled sand can be measured within a few metres. In this particular instance, little movement was observed over three months, despite some significant storm activity.

Coastal engineering demonstrations

The technique has been widely applied to coastal engineering investigations, examples include the following.

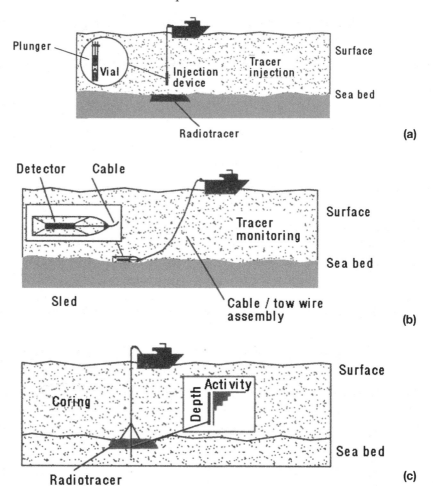

Figure 9.5.   Application of tracer techniques to sediment migration. (a) Injection of
the tracer using a bottle breaker technique. For depths greater than, say, 20 m a
detonator technique for dispersing the tracer has advantages. (b) The tracking of the
labelled sediment for periods ranging from a few days to many months. (c) The depth
distribution of the tracer is best obtained by coring. (d) Case study showing that sand
moves very slowly at a depth of 68 m over a period of over 3 months. (e) Case study
showing that the direction of bed load transport can vary across a channel depending
on the net effect of the current patterns. (f) Schematic representation of nucleonic
sediment gauges; a transmission gauge (left) and a backscatter gauge (right).

- *Shipping channels:* bed load transport studies may be used to optimise the
  alignment of dredged shipping channels. For instance, tracer studies have been
  used to measure the rate of littoral drift which will lead to in filling of the proposed
  channel and the need for ongoing maintenance.
- *Location of dredge spoil grounds:* optimising the location of dredge spoil grounds

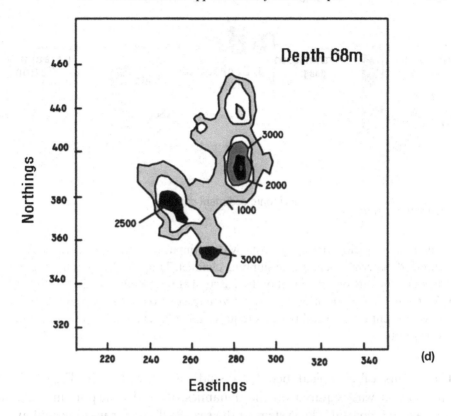

(d)

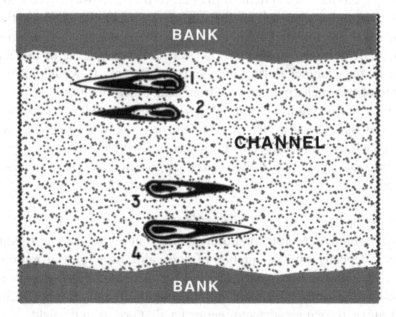

(e)

Figure 9.5.   (*cont.*)

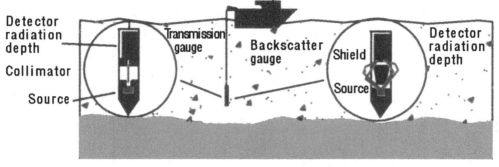

Figure 9.5.   (*cont.*)   **Nucleonic sediment gauges**   (f)

involves balancing dredging costs with estimates of the efficiency with which dumped material re-enters the shipping channel. Tracer techniques have been used to study the rate of movement of dredge spoil at proposed dump sites.

• *Sustainable development in the coastal zone:* prediction of the impact of a proposed development on the sand budgets requires good baseline information on bed load transport.

The results of a typical bed load study are sketched in Figure 9.5(d). Information was required on the dynamics of sand transport in a channel connecting a coastal lake system to the sea. Radiotracer was injected at four locations across the channel. In the northern part of the channel, the net migration was in one direction; in the southern section, in the opposite direction. The observed sand migration was the consequence of the current patterns in the channel during ebb and flood tides. It would have been difficult to obtain such direct information using other methods.

A number of case studies in Latin America and Europe have been reviewed by Aun and Bandiera in IAEA (1995). On the basis of long experience, the authors concluded that the benefits of tracer technology will only be fully realised as part of a multi-disciplinary approach to coastal engineering problems.

Detector calibration

Success in obtaining reliable quantitative information on sand and sediment movement depends on the quality of the calibration of the monitoring detector. As with the total count method for measuring flow rates in rivers and pipelines (Sections 9.3.2 and 8.3.6), the major difficulty is reproducing, in the laboratory, the source–detector geometry that occurs in the field.

Calibration of the detector efficiency is made in a large water tank partially filled with sand and designed to simulate conditions at the bottom of the lake or estuary. A realistic geometry can often be simulated by mounting the detector on a rig a few centimetres above the surface of the sand (Figure 9.2(b)).

Complications can arise in the field because the radioactive sand released to the surface is incorporated into the mobile layer as a consequence of periodic re-suspension and deposition processes associated with bed load transport. Over extended periods, bioturbation might also be significant. As a consequence of γ ray attenuation, the efficiency with which the detector responds to radioactive sand particles depends on its depth. This effect may be simulated in the calibration tank by covering the radionuclide source with successive layers of sediment and taking repeated measurements. When calculating the calibration factor an assumption has to be made concerning the distribution of the tracer as a function of depth. The validity of the assumption should be checked if possible by collecting cores at a number of locations within the radioactive patch, sectioning the cores and measuring the activity profiles directly (Figure 9.5(c)).

Suspended sediment gauges

Nucleonic gauges have been developed for the accurate measurements in real time of the levels of suspended sediments in rivers and estuaries (Figure 9.5(f)). These data are of interest in understanding issues as diverse as the stability of dredge spoil grounds and, more generally, the impact on estuaries of sediments transported to the coastal zone by river systems. Suspended sediment gauges are designed as gamma ray transmission (Section 7.2.1) or gamma ray backscatter (Section 7.2.2) gauges. Both types of gauges can measure sediment loadings equivalent to bulk density changes of less than 1%. With transmission gauges, the sensitive region is the small volume between the source and the detector. This is a disadvantage compared with the backscatter gauge, which has a much larger sensitive volume. On the other hand, the response of the transmission gauge is much less sensitive to effects due to the proximity to boundaries such as the bed of the estuary or the water surface. Applications have been made to:

- continuous monitoring of high (say 1 to 100 kg/m^3) sediment loadings in rivers as a function of rainfall and at different depths;
- surveying the levels of suspended sediments in reservoirs and harbours as a function of location and depth; and
- optimising the efficiency suction dredging by monitoring the concentration of dredge spoil in the well of the dredge as a function time.

Further information may be obtained from Tazioli and Caillot (IAEA, 1983) and Aun and Bandeira (IAEA, 1995).

9.4 Applications of naturally occurring radioisotopes

9.4.1 Man-made versus environmental radioisotopes

Naturally occurring, or environmental radioisotopes were introduced in Section 9.2.2 when they were classified into three categories: cosmogenic isotopes, fall-out products and primordial isotopes (Table 9.2). Applications of man-made and environmental isotopes are generally complementary. Man-made radionuclides are selected from a large number of available isotopes, are used to label a particular component of a complex system and are injected at a precise location and according to a precise protocol. Consequently, very detailed information is obtained on the behaviour of an environmental system in the vicinity of the injection.

By contrast, environmental isotopes are widely distributed by natural processes and are used to obtain information over a regional scale. By and large the data reflect the cumulative effect of the environmental process (say, erosion, siltation, groundwater movement) over time scales related to the half life of the isotope. Information is normally obtainable over periods equivalent to four or five half lives. For instance ^{210}Pb ($T_{1/2} = 22.3$ y) is used to assess siltation rates over about 100 years, whereas, ^{14}C ($T_{1/2} = 5730$ y) and ^{230}Th ($T_{1/2} = 75\,400$ y) provide information over about 20 000 and 200 000 years respectively.

9.4.2 Erosion studies

A new use for caesium-137

Good husbandry of fertile soil is central to sustainable development in a world where populations are expanding. Every effort is being made to minimise losses due to water and wind induced erosion. Part of this effort involves the development of reliable measurement techniques. A great deal is being achieved with satellite imaging. However, there is a need to validate or interpret the imaging data with independent estimates of erosion patterns at a catchment scale. One such method involves the use of the environmental isotope ^{137}Cs to estimate erosion rates over a whole catchment area. ^{137}Cs is a fission product which was injected into the upper troposphere and lower stratosphere during atmospheric nuclear tests. These radionuclides were distributed world-wide by the global atmospheric circulation and, with the passage of time, diffused to the surface of the earth. There the ^{137}Cs adsorbed strongly onto the clay fraction of the soils, and acted as a natural radiotracer for subsequent erosion and accumulation processes.

Measurable levels of ^{137}Cs started to appear in soil during the mid-1950s, reached a peak in the mid-1960s and then declined following the signing of the Atmospheric Test Ban Treaty in 1964 (Playford *et al.*, 1992). The year by year variation of the levels of ^{137}Cs reaching the earth is similar to that of other 'fall-out' products such as tritium and carbon-14 (Figure 9.6(a)) and is known as the input function. Caesium-137 has a sufficiently long half life (30.1 y) to remain useful for the present purpose for many more years to come.

Environmental ^{137}Cs was first used as a natural tracer to measure the rate of sedimentation in lakes and reservoirs (Pennington, 1972, Pennington *et al.*, 1973). The sediment with the ^{137}Cs attached accumulates in the lake as a consequence of erosion in the surrounding catchment areas (Figure 9.6(b)). If there is normal sheet erosion over the catchment, the shape of ^{137}Cs input function is preserved in the sediments. It is the shape of the profile and not the radionuclide decay which serves as the basis for dating the accumulation of the sediments. The first appearance of environmental ^{137}Cs in the mid-1950s and the peak in the mid-1960s are both readily identifiable features. These data may be compared with results from other dating methods and with the known historical record to gain an improved understanding of changes in erosion patterns.

Experimentally, the ^{137}Cs profile is measured by

- collecting and sectioning a core obtained from the soil in the region of interest;
- placing each dried and weighed section of the core in a Marinelli beaker over a high-resolution detector (Figure 9.6(c));
- measuring the countrate in the 662 keV γ ray peak; and
- calculating the specific activity of the ^{137}Cs by comparison with results obtained from a calibrated standard.

Procedures and applications

The measurement, and in particular the prediction, of the rate of erosion of fertile soils is a key to the development of strategies for the long-term sustainability of agriculture. The rate of surface erosion depends on a number of parameters such as the distribution, the amount and the intensity of rainfall, the slope of the land, the nature and extent of vegetation and the erodability of the soil. Attempts have been made to develop equations linking erosion rates to these parameters, and to use the equations to predict erosion over a wide geographic region (Ritchie and Ritchie in IAEA, 1995). One of

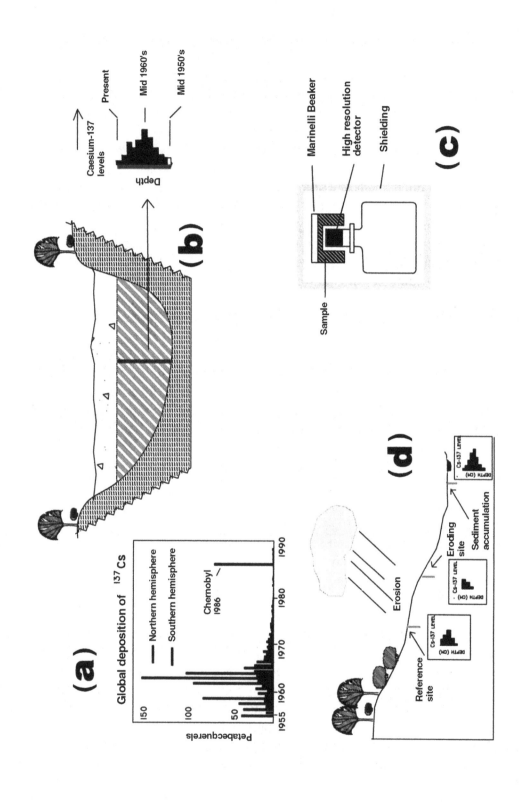

(a)

Global deposition of ^{137}Cs

— Northern hemisphere
— Southern hemisphere

Chernobyl 1986

Petabecquerels

150
100
50

1955 1960 1970 1980 1990

(b)

Caesium-137 levels

Present
Mid 1960's
Mid 1950's

Depth

(c)

Marinelli Beaker
High resolution detector
Shielding

Sample

(d)

Erosion

Eroding site

Sediment accumulation

Reference site

Cs-137 LEVEL

DEPTH (CM)

the most widely used is the Universal Soil Loss Equation (USLE). However, it has often been inappropriately applied, both in the United States where it was developed and around the world (Wischmeier, 1976). An independent method for the validation of such equations is required.

Carefully planned and executed measurements of the ^{137}Cs profiles on the catchment slopes provide a basis for independent estimates of erosion rates. The principle is shown schematically in Figure 9.6(d). The greater the erosion rate, the greater the removal rate of soil and the less the residual ^{137}Cs in the soil profile. On the other hand, the greater the rate of accumulation of soil in a catchment, the greater the level of ^{137}Cs in its profile. Activity profiles similar to those sketched in the figure are not difficult to observe. A key for the reliable interpretation of the resulting data is the identification of a reference location on the slope where there has been minimal erosion or sedimentation over the past decades. The identification of the reference site usually relies on the judgement of local experts.

A complicating factor is that the countrate in the 662 keV peak attributable to ^{137}Cs is often little above background. Experimentally, the background may be reduced by using a shielded gamma spectrometer with a Compton suppression capability. Such equipment is, however, specialised and expensive. Mathematical procedures for calculating the specific activities of gamma emitting isotopes at very low levels in natural samples are available, but their discussion is beyond the scope of this text (Gilmore and Hemmingway, 1995).

An interesting result

The application of ^{137}Cs to the study of erosion has proven useful to studies not only on a regional scale but also at the level of an individual farm and in forested areas. Measurements can contribute to studies of the impact of land use management on erosion. For instance, a study on a vineyard revealed that for each bottle of wine produced, approximately one bottle of fertile soil was lost to the cultivated area. If sustainable development is to be achieved, an understanding is needed of the total environmental cost of production.

Other techniques

The procedure for using ^{137}Cs as a dating technique is different from that used for the majority of radiometric methods. It does not depend on the

◀ Figure 9.6. Application of environmental ^{137}Cs to erosion measurements. (a) The global distribution of ^{137}Cs (Playford *et al.*, 1992). (b) The dating of sediment accumulation in lakes and reservoirs from ^{137}Cs levels in drill core. (c) Counting of sediment samples in a Marinelli beaker; (d) Application of ^{137}Cs measurements to catchment erosion.

measurement of radioactive decay but on identifying special features of the input function such as the onset of the ^{137}Cs activity (mid-1950s) and the activity maximum (mid-1960s).

There are also the conventional methods of radiodating using ^{210}Pb ($T_{1/2} = 22.2$ y),^{14}C ($T_{1/2} = 5730$ y) or ^{230}Th/ ^{234}U (Figure 1.5), which can serve to measure sediment accumulation over time scales up to decades, tens of thousands and hundreds of thousands of years respectively. In these ways, insights can be obtained on the evolution of estuaries or other geologic features over recent geological time. Some details are presented in Table 9.2.

9.4.3 Groundwater

Introduction

Adequate resources of fertile soil and clean drinking water are essential for public health. To complement the discussion of the measurement of erosion offered in the previous section, attention will now be paid to studies of sub-surface water. This resource becomes increasingly important in areas where reserves of good quality surface water have diminished due, for example, to prolonged dry spells, over-exploitation or contamination. The challenge is to establish management regimes that will ensure the sustainability of ground-water yields.

Groundwater resource evaluation

Environmental radioisotopes have proved a useful aid in the investigation of underground water with a few examples listed in Table 9.1. Applications have been directed towards:

- identifying the sources of recharge water, i.e. the sources of surface water which seep underground to replenish the basin;
- estimating the extent of mixing of underground water from different sources;
- calculating the age of groundwater samples, i.e. the time which has elapsed since the water percolated underground;
- deducing the direction and rate of groundwater flow; and
- understanding the processes leading to degradation in groundwater quality.

The environmental isotope approach is here illustrated with reference to a simplified version of the water table (Figure 9.7). The data obtained for this purpose are often used to validate groundwater hydraulic models which link the known water table levels (technically the piezometric surfaces) to the properties of the water bearing strata (Clark and Fritz, 1997).

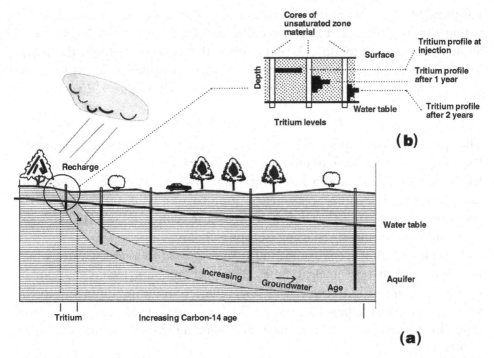

Figure 9.7. Application of environmental isotopes to groundwater studies. (a) Schematic diagram showing underground water percolating through an aquifer and sampled at various locations for age dating with, say, carbon-14. (b) The use of man-made tritium to study the rate of recharge of infiltrating rainwater to an aquifer.

Two examples will be provided: the location of recharge areas and the dating of underground water.

Location of recharge areas

Recharge is the process whereby surface water percolates through the soils and other overlying strata to replenish the underground water. It is therefore important to ensure that areas of recharge are clearly identified and well managed. Environmental tritium can make a significant contribution to this end. As shown schematically in Figure 9.7(a) the direction of groundwater flow is away from recharge areas. The presence of significant levels of tritium is indicative of groundwater which has infiltrated through the unsaturated zone in post-nuclear times (Table 9.2), that is of samples collected from a bore in or close to the recharge area. It is recalled that most of the tritium in the environment was derived from atmospheric nuclear testing.

As an aside, it may be noted that small injections of man-made tritium can be used to obtain detailed information on the transport of water in the unsaturated zone. Tritium is injected at a constant depth (say 1 m). Cores within the unsaturated region are collected at time intervals ranging from about 1 month up to, say, 3 years. With the passage of time, the tracer maximum is displaced down the profile towards the water table. At the same time, it exhibits dispersion. This behaviour is shown in Figure 9.7(b).

The dating of underground water

The age of a groundwater sample is the time required for the water to migrate from the area of recharge to the extraction bore. Groundwater migrates away from the recharge area at a rate which depends on differences in the water table levels (the so-called piezometric gradient) and the properties of the water bearing strata. Groundwater ages can either be computed from hydraulic data, or measured using isotope techniques.

As indicated above, environmental tritium is used to identify water which has recharged in post-nuclear times. However, modern techniques can be used to measure tritium to such low levels that groundwater dating up to 100 years is possible. Carbon-14 ($T_{1/2} = 5730$ y) is widely used for the dating of groundwater samples with residence times ranging from a few hundred up to 30 000 years.

Unlike tritium (^3H), carbon is not a component of water but a solute. The calculation of the groundwater age from the measured activities involves not only an assessment of the carbon-14 levels at recharge, but also an under-standing of the source and subsequent behaviour of the dissolved carbonate in the groundwater system. By dating a large number of samples across a basin, contours of equal age (isochrones) can be drawn from which the groundwater flow lines can be deduced.

Chlorine-36 ($T_{1/2} = 302\,000$ y) may be used to date water with residence times in excess of one million years. For instance, the Great Artesian Basin, Australia, covers about a third of the continent. The major recharge area is in the east, and the water flows in a generally westerly and southwesterly direction. The size of the basin is such that groundwater ages reach two million years near some of the discharge areas. In order to validate chlorine-36 as a groundwater dating method, an inter-comparison was made between the isotopic ages, and those calculated from a hydraulic model. Satisfactory agreement was found.

The atom ratio $^{36}Cl/Cl$ depends not only on the age and the hydraulic properties of the groundwater, but also on the extent to which it has dissolved sub-surface chloride. Chlorine-36 techniques are being widely used to study salination and the degradation of groundwater quality.

Accelerator mass spectrometry

The carbon-14 and chlorine-36 dating methods make use of Eqs. (1.5) and (1.7) (Section 1.6.1), i.e. $dN/dt = \lambda N = 0.693\ N/T_{1/2}$ where, N refers to the number of ^{14}C atoms in the sample, while λ and $T_{1/2}$ refer to the ^{14}C decay constant and half life respectively. The accelerator mass spectrometry (AMS) technique which is normally built into a tandem accelerator measures the number of atoms N directly. It is therefore well suited to long half life nuclides and so not only for ^{14}C and ^{36}Cl but also for ^{10}Be, ^{26}Al, ^{41}Ca and many others (Table 9.2). The nuclides from the target are accelerated in a two-stage process to the MeV region when they enter a strong magnetic field which deflects them into multiple detector systems. AMS is used to measure very low-abundance isotopes on small samples. For instance, only milligram quantities of chloride in silver chloride or graphite are needed to measure ^{36}Cl/Cl or ^{14}C/C to atom fractions of 10^{-15}. Technically, the central challenge is to design a detection system which distinguishes the isotope of interest from that of much more abundant impurity isotopes of the same mass number (i.e. to measure ^{36}Cl in the presence of residual ^{36}S or ^{14}C in the presence of residual ^{14}N).

Accelerator mass spectrometry has been widely applied to archaeology, glaciology, geochemistry and oceanography. Readers interested in AMS and its applications are referred to the excellent book by Tuniz *et al.* (1998).

9.4.4 Oceanography

Environmental isotope techniques are well suited to the study of the dynamics of large and complex systems and have contributed significantly to the understanding of many characteristics of the oceans. There are two classes of investigations. The first concerns the processes within the coastal zone resulting from, say, the impact of river systems leading to the release of contaminants derived from agriculture, industrial development or urbanisation. The second concerns circulation patterns within the deep oceans and their impact on weather and climate change and on fisheries.

There are two principal groups of radionuclides in the oceans, though both of them are present in only very low concentrations. The first includes the isotopes tritium and carbon-14 which enter the ocean in rainfall and by exchange with the ^{14}CO$_2$ of the atmosphere respectively. The second source is thorium and uranium and their radioactive daughters (Figure 1.5) which enter the coastal zone dissolved in river water or associated with river sediments.

Extensive surveys have been made over the years to estimate tritium and ^{14}C concentrations. Tritium is particularly useful in the study of mixing processes in the upper ocean. A fairly complete understanding of oceanic circulation patterns has been obtained from the study of large numbers of profiles. Both tritium and carbon-14 exhibit a 'bomb pulse' (Section 9.3). The carbon-14 bomb pulse is detected down to about 1000 m. Below that depth the cosmogenic source dominates.

Systematic study of the profiles leads to insights into the oceanic circulation patterns. Generally speaking, below a few hundred metres, the ^{14}C levels decrease with depth. However, upwelling of deep ocean water which contains very little ^{14}C leads to lower than normal values at the surface. Carbon-14 levels effectively act as a tracer for these packets of deep ocean water.

Such data associated with much other information have led to the 'conveyor belt' picture of global circulation patterns. Cold water is conveyed along the ocean floor from the North Atlantic via the Antarctic region to the North Pacific. The return water comprises warm surface water from the Pacific via the Indian Ocean to the Atlantic.

9.5 Nuclear waste disposal

9.5.1 The need for complete isolation

Nuclear power production leads inevitably to highly radioactive waste products which include a mixture of highly radioactive fission products and actinides together with some activation products. The volume of this waste is much less than 1% of the waste generated by coal fired power stations per 1000 MW of electrical power. However, the highly radioactive waste has to be isolated from contact with the biosphere, for thousands to tens of thousands of years or longer. Furthermore the integrity of the disposal concept must be guaranteed with the highest possible certainty (ICRP, 1997).

To effect isolation, the nuclear industry uses a defence in depth philosophy. The concept is based on establishing multiple barriers between the radioactive material and the biosphere, including:

- immobilisation of the fission products and actinides in a solid matrix;
- encasing the immobilised material in specially designed casks;
- placing the casks in engineered repositories and backfilling with bentonite or similar material; and
- siting the installation so as to ensure that any radioactive waste products which might be leached from the repository are as far as possible immobilised in the surrounding geological strata.

9.5.2 *Natural analogues*

Multi-barrier systems

Attention is paid to the optimum selection of a repository site to ensure that the geological conditions of the burial site contribute to the immobilisation of waste products to the maximum extent possible.

Procedures to obtain official approval for radioactive waste repositories are beset with a particularly serious problem. As indicated above, proponents are required to convince regulators that the repository will meet specifications for very long periods into the future. Every element of the multi-barrier system comprising the repository system must be analysed with utmost care. An important component of the multi-barrier system is the geosphere around the proposed repository. Long-term predictions of radionuclide retardation in the geosphere are based on numerical models, the validation of which is clearly of the utmost importance.

An approach to the difficult problem of validating transport codes for long-term predictions is the use of natural analogues. A natural analogue is a feature in the environment exhibiting processes which are similar in essential respects to those being modelled. A wide range of examples have been discussed in Miller *et al.* (1994). A number of uranium ore bodies have proven useful as natural analogues of repositories for spent nuclear fuel.

A natural analogue of the leaching of fission products from spent fuel

The Oklo uranium deposit in the country of Gabon in Central Africa is one of the best known natural analogues for this purpose. It contains several sites where chain reactions in uranium occurred naturally and repeatedly over periods of a couple of hundred thousand years to some two to three billion years ago when ^{238}U levels were some 30% and the ^{235}U some 6.8 times higher than today.

These sites comprise a unique natural analogue of the fate of spent reactor fuel. During the long active period about 5.4 tonnes of fission products were generated as well as 1.5 tonnes of plutonium and other transuranic elements (Hore-Lacy, 1999, Section 5.4). Although the fission products have long since decayed, the isotopic signatures of the stable end products remain. From a detailed analysis of their distribution within the host rock, it is possible to conclude that some of the fission products and their decay products were bound in the host rock over this very long period of time. This applies even though groundwater had ready access to the geological formations for long periods of time.

The analogue approach first explored at Oklo has been extended to many normal uranium deposits. In most studies, measurements are made in the surrounding environment of radionuclides leached from the deposit by infiltrating groundwater. The distribution of such radionuclides is an analogue of the distribution of similar waste products which would be leached from spent reactor fuel if the waste repository were breached by groundwater.

Many natural analogues of various components of a waste repository system have been studied. These include analogues of the immobilising medium (boro-silicate glass or the mineralogical components of synthetic rock (synroc)), of the cask material (iron or copper) and of the backfill clay materials. They do not involve the use of radiotracers and are therefore outside the scope of this book. It suffices to say that natural analogue studies have generally added to the confidence that nuclear material may be safely stored for very long periods of time in properly designed facilities.

9.5.3 *Regulatory requirements*

These examples have demonstrated that natural analogues simulating the processes of interest here are extremely helpful in enhancing confidence in various concepts employed for planning the long-term storage of radioactive waste. However, regulators also look for a more systematic approach and require proponents of repository designs to undertake detailed performance assessments. Such assessments are designed to provide an assurance that all regulatory requirements are met. They involve the mathematical modelling of many aspects of repository performance, including the behaviour of radionuclides which would be leached into the surroundings should the repository be breached by groundwater.

9.6 Summary and conclusions

Both, man-made and environmental radioisotopes have long been widely applied to environmental studies. Their usefulness is principally due to the fact that individual components of complex systems can be separately labelled and then separately studied. Not only are a wide range of radiotracers available for tagging specific materials, but the tracers can often be detected in the positions they naturally occupy and this with high sensitivity. Several examples were discussed, including the measurement of river flow rates, the offshore dispersion of sewage and other contaminants and the migration of sand and sediments.

Environmental isotopes are those which are very widely dispersed through a range of ecosystems. They were here classified into three groups according to their sources, namely, radioisotopes produced by atmospheric nuclear testing, cosmogenic isotopes and primordial isotopes.

It was also emphasised that environmental isotopes are applied in two conceptually different ways. First, they are used as dispersed radiotracers. Examples which were quoted included the application of ^{137}Cs to the study of erosion and sediment redistribution and the use of tritium in the mapping of the recharge areas to groundwater systems. Second, environmental isotopes are used to date selected materials such as sediments from cores in lakes and estuaries and water from bores along flow lines in groundwater basins. A special case involves the use of uranium series disequilibria to study the migration of uranium and its daughter products in the vicinity of uranium ore bodies since such ore bodies may be used as natural analogues of high-level nuclear waste repositories.

Much environmental work is designed to support ecologically sustainable development. The implementation of such a policy requires the assessment of the future impact of contemporary development proposals. Two approaches have been discussed. The first involves the use of predictive mathematical models followed by applications of tracer techniques to validate the code. The second approach uses dating techniques to establish archival records of environmental processes which form a basis for prediction.

To conclude, we hope to have demonstrated what we claimed in our foreword. Radioisotopes and the emitted radiations serve as tracers and environmental monitors for a very large number of applications using only relatively standard equipment. They are tools which can be expected to behave reliably and efficiently for simple as well as complex tasks in applied science and engineering, so contributing to the general wellbeing.

Appendix 1

Glossary of technical terms

This glossary lists terms in common use in the nuclear sciences. Fifteen terms used in radiation protection are listed in Table 2.2 and properties of nuclear particles are summarised in Table 1.2.

Absorption   (1) The retention, in a material, of energy removed from radiation passing through it. (2) The process whereby neutrons (or other elementary particles) are captured by nuclei.

Activation analysis   A method of analysis based on the identification and measurement of characteristic radiation from radionuclides formed by neutron irradiation in a sample of the irradiated material.

Activity (of a substance)   Short for radioactivity. The number of disintegrations per unit time taking place in a radioactive material. The unit of activity is the becquerel (Bq), one disintegration per second.

Alpha particles   Energetic helium nuclei (^{4}He) emitted from the nuclei of higher Z number atoms ($Z > 60$) during alpha particle decays.

Annihilation radiation   The electromagnetic radiation resulting from the mutual annihilation of the mass of negatively charged electrons (negatrons) and positrons ($\beta^- - \beta^+$). The annihilation radiation consists of two photons, each 0.51 MeV, emitted in opposite directions.

Atomic mass unit   One-twelfth of the mass of a carbon-12 atom and approximately equal to the mass of a single proton or neutron. It is abbreviated amu.

Atomic number   The number (Z) of protons in the nuclei of the atoms of an element, indicating the position of that element in the periodic table.

Atomic weight   For a given specimen of an element, the mean weight of its atoms, expressed in either atomic mass units (physical scale) or atomic weight units (chemical scale).

Auger electrons   Electrons emitted from atoms (in competition to X rays), using the energy released following the filling of vacancies in inner electron shells.

Background radiation   The ionising radiation in the environment to which everyone is exposed. It comes from outer space, the sun, terrestrial rocks and soil, from buildings, the air that is breathed, the food that is eaten and from human and animal bodies.

Barn   A unit of cross section. 1 barn = 10^{-28} m^2; 1 millibarn = 10^{-31} m^2.

Becquerel   Unit of activity, abbreviated Bq, equal to one radioactive disintegration per second. Replaces the curie (Ci): 1 Ci = 3.7×10^{10} Bq.

Beta particles   Electrons, of positive or negative charge, emitted from the nuclei of atoms during β particle decays. (*see also* electrons)

Bremsstrahlung   The German word for braking radiation. The electromagnetic radiation resulting from the retardation of charged particles travelling through matter.

Carrier-free   A preparation of a radioisotope to which no non-radioactive carrier has been added. Material of high specific activity is often loosely referred to as 'carrier-free'.

Contamination   A deposit of dispersed radioactive material on or within any other medium, either solid, liquid or gaseous.

Cross section   A measure of the probability of a particular nuclear reaction between a projectile and a target. The probability is expressed as the area that the target presents. The unit is the barn, 1 barn = 10^{-28} m^2.

Curie   Abbreviated Ci and historically the first unit for the measurement of radioactivity. Superseded by the becquerel (Bq). It is now defined as 1 Ci = 3.7×10^{10} Bq.

Cyclotron   A machine to accelerate charged atomic particles to megaelectronvolt energies by the application of electromagnetic forces. These particles are often used to bombard target materials to produce radionuclides.

Daughter   A nuclide or radionuclide that originates from a radionuclide (known as the parent) by radioactive decay.

Decay, radioactive   A spontaneously occurring change in an atomic nucleus commonly proceeding via α or β particle emission or electron capture (EC). The large majority of these decays, as they are called, are followed by secondary or follow-on radiations. (*see also* secondary radiations)

Deuterium   Also called 'heavy hydrogen'. A non-radioactive isotope of hydrogen having one proton and one neutron in its nucleus instead of just one proton. Its natural abundance is 1 to 6500 hydrogen atoms.

Dosemeter   A device, such as a film badge, to measure the radiation dose a person receives over a period of time.

Electrons   Elementary particles, each carrying a negative charge of 1.6022×10^{-19} C, either negative (negatrons) or positive (positrons). Negatively charged electrons are associated with each neutral atom, their negative charge balancing the positive charge of the protons. (*see also* beta particles)

Electron capture decay (EC)   A nuclear transformation whereby a nucleus captures one of its orbital electrons normally from an inner electron shell.

Electronvolt (eV)   The kinetic energy acquired by an electron when accelerated through a potential difference of 1 volt. 1 eV = 1.6022×10^{-19} J.

Element   Matter consisting of atoms having the same atomic number.

Excitation   In nuclear sciences, the addition of energy to a system, following a nuclear transformation, moving the system from its ground state to an excited state. The inverse is known as de-excitation.

Fast neutrons   Commonly defined as neutrons of energy exceeding 10 keV. (*see also* neutrons)

Film badge   One or more photographic films and appropriate filters (absorbers) used for the measurement of radiation exposure or quantities related to absorbed dose. It is commonly worn by radiation workers.

Fission   In the nuclear sciences it is the splitting, by neutron interaction, of a heavy nucleus into two radioactive 'daughters' of nearly equal masses. Fission is promptly followed by decays of the 'daughters' with the emission of neutrons, electrons and gamma radiation, the emitted energies being of order 10^8 eV per fission.

Fission products   The stable and unstable nuclides resulting from fission, including neutrons which can cause a chain of additional fissions.

Fluorescent X rays   Electromagnetic radiations emitted from atoms (in competition with Auger electrons) using the energy liberated by the filling of vacancies in inner electron shells. (*see also* Auger electrons, X rays)

Flux, neutron   The number of neutrons passing through unit area in unit time.

Follow-on radiations   See secondary radiations.

Fusion   The formation of heavier nuclei following the fusing of two or more lighter nuclei (e.g. hydrogen atoms) with an attendant release of energy, usually of order 10^8 eV per fusion reaction.

Gamma radiation   Electromagnetic radiation emitted by atomic nuclei.

Ground state   The state of lowest energy of a system, usually of an atom.

Half-life   The time in which the activity of a radionuclide decays to half its initial value.

Half-life, biological   The time required for the amount of a radioactive substance in a biological system to be reduced to one half of its initial value by both, radioactive decays and biological processes.

Heavy water   Water containing significantly more than the natural proportion (1 in 6500) of heavy hydrogen (deuterium) atoms to ordinary hydrogen atoms. Heavy water is used as a moderator in fission reactors.

High-level waste   Radioactive material for which no further use is contemplated. High level waste emits sufficiently large levels of ionising radiation and heat to require the maximum standards of radiation protection when the waste is handled, transported or stored.

Internal conversion (IC)   The de-excitation of a nucleus in which the liberated energy is transferred to an orbital electron (instead of emitted as a γ ray) which is thereby ejected from the atom.

Internal conversion coefficient   The ratio of internal conversions (IC) following a nuclear decay to the number of de-excitations by γ rays.

Ions   Atoms that have lost or gained one or more orbiting electrons, thus becoming electrically charged.

Ionisation   Any process by which atoms, molecules, or ions gain or lose electrons.

Ionising radiation   Radiation capable of causing ionisation of the atoms through which it travels.

Irradiation   This occurs when ionising radiation impinge on material.

Isomeric state   State of a nucleus above its ground state which has a long-enough half life to be used for experimentations.

Isomeric transitions   The decay of isomeric or metastable states to their ground state, abbreviated IT.

keV   Thousand electronvolts = 10^3 eV.

Labelling, radioactive   Incorporation of a radioactive tracer into a molecular species or macroscopic sample when the emitted radiation facilitates detection.

Low-level waste   Radioactive material for which no further use is intended. Such materials contain only low levels of radioactivity, low enough to be adequately safely stored by simple burial in locations which are clearly marked.

LS counting   Acronym for liquid scintillation counting. A method utilising scintillators dissolved in an organic liquid for the absorption of ionising radiations. The scintillator generates light quanta which are amplified by a photo-multiplier (PM) tube and displayed by signal processing equipment.

Mass number A number assigned to a nuclide. It is an integer, equalling the number of protons plus neutrons in its nucleus.

Metastable states In a nucleus, an alternative term to isomeric states.

MeV Million electronvolts = 10^6 eV = 10^3 keV.

Moderator A material, such as ordinary water (H_2O) or heavy water (D_2O) or graphite, commonly used in a fission reactor to slow down high-energy neutrons thus increasing the probability of further neutron reactions.

Negatrons Negatively charged electrons. (*see also* beta particles)

Neutron Accounts, together with protons, for the particles in the nuclei of atoms. Neutrons are of zero charge and 1.0014 times as heavy as protons. (*see also* thermal neutrons)

Nucleus The positively charged central portion of an atom consisting of protons and neutrons with which is associated 99.95% of the atom's mass but it occupies only 10^{-12} of its volume.

Nuclide An alternative term to isotope.

Photon A quantum of electromagnetic radiation of energy $h\nu$ (h being Planck's constant and ν the frequency of the radiation).

Positrons Positively charged electrons. (*see also* electrons)

Primary radiations Radiations which signal transformations of parent nuclides to daughter nuclides with a corresponding change in their Z number. Examples are α or β radiations.

Proton The nucleus of a hydrogen atom having a charge equal and opposite to that of an electron and a mass of 1.0076 amu.

Radiation source A quantity of radioactive material used as a source of ionising radiation.

Radioactivity concentration The activity per unit quantity (mass or volume) of any radionuclide.

Radiochemical purity Of a radioactive material, the proportion of the total activity that is present in the stated chemical form.

Radionuclide An alternative term to radioisotope. (*see also* nuclide)

Radionuclidic purity The degree, high or low, to which radioactive material is present in the stated chemical form.

Radiotracer A radioisotope introduced into a system to investigate its behaviour with the help of the emitted radiations.

Scintillation A process occurring in selected materials known as scintillators when detectable light quanta are generated by ionising radiations.

Scintillation detector A device which detects ionising radiations by using their ability to produce light quanta.

Secondary radiations Also known as follow-on radiations. They are emitted from the daughter which are created in an excited state due to a radioactive decay. The daughter de-excites, usually with the emission of γ rays and conversion electrons, but also fluorescent X rays and Auger electrons.

Specific activity The activity per unit mass of an element or compound containing a radioactive nuclide.

Thermal neutrons Neutrons in thermal equilibrium with their surroundings. At room temperature their mean energy is about 0.025 eV. Neutrons of energies above 10 keV are known as fast neutrons. (*see also* neutrons)

Tritium A radioisotope of hydrogen (^{3}H). It is a very rare component of the atmosphere regenerated at high altitudes by the reaction of fast nuclei with nitrogen $^{14}N(n,^{3}H)^{12}C$.

X rays Electromagnetic radiations of energies exceeding about 0.1 keV but there is no upper limit. They are produced as bremsstrahlung or as fluorescent radiations.

Appendix 2

A selection of texts on health physics and radiation protection

These titles were compiled from the list of References at the end of Chapter 9. They are brought together here for easy access. The first ten entries are textbooks, followed by references to three publications in peer-reviewed journals.

Cember H. (1996) *Introduction to Health Physics*, 3rd edn. McGraw-Hill, New York, USA.

IAEA (1986) *Radiation Protection Glossary*, IAEA Safety Series No.76, International Atomic Energy Agency, Vienna, Austria.

IAEA (1996a) *Basic Safety Standards for Protection Against Ionising Radiation and for the Safety of Radioactive Sources*, IAEA Safety Series No.115, International Atomic Energy Agency, Vienna, Austria.

ICRP (1991) *Publication 60*, 1990 Recommendations of the International Commission on Radiological Protection, Vol 21, pp. 1–3, Pergamon Press, Oxford, UK.

ICRU (1993) *Quantities and Units in Radiation Protection Dosimetry*, ICRU Report 51, International Commission on Radiation Units and Measurements, Bethesda, Maryland, USA.

Martin A. and Harbison S.A. (1996) *An Introduction to Radiation Protection*, 4th edn, Chapman and Hall, London, UK.

Simmons J.A. and Watt D.E. (1999) *Radiation Protection: A Radical Reappraisal*, Medical Physics Publishing, Madison, Wisconsin, USA.

Shapiro J. (1981) *Radiation Protection, A Guide for Scientists and Physicians*, 2nd edn, Harvard University Press, Cambridge, Mass, USA.

Shleien B. (Ed) (1992) *The Health Physics and Radiological Health Handbook*, Revised Edition, (Scinta Inc, Silver Spring MD 20902, USA). See also Shleien, B, Slaback, L.A. Jr. and Birky, B. (Eds), (1997), 3rd Edition, William and Wilkins, POB 23291, Baltimore, MD 21298-9533, USA.

Turner J.E. (1986) *Atoms, Radiations and Radiation Protection*, Pergamon Press, New York, USA.

Clarke R.H. (1991) The Causes and Consequences of Human Exposure to Ionising Radiations, *Radiation Protection Dosimetry*, **36**, 73–77.

Graham *et al.* (1999) *Low Doses of Ionising Radiations Incurred at Low Dose Rates*. A conference paper prepared on behalf of the International Nuclear Societies Council. *Radiation Protection in Australasia*, **16**(1), 32–47.

Luckey T.D. (1998) *Impressions of the IAEA/WHO Conference on Low Doses from Ionising Radiations*, Nov. 1997, Seville, Spain. *Radiation Protection Management*, **15**, 1, Guest Editorial.

Appendix 3

Comments on the availability of nuclear data on the Internet

Electronic data exchange has become routine. Large research institutes are making their data widely available via the Internet as a cost-free service to the scientific community.

Nuclear reference data are particularly extensive and well suited for electronic distribution, as was noted in Section 4.2.1 where a brief reference was made to data from the web site of the Nuclear Data Center of the US Brookhaven National Laboratory (www.nndc.bnl.gov). The information at present available to users is the product of the combined efforts of the US National Nuclear Data Centre (NNDC) with other data centres and other interested groups which have an interest in such data, not only in the United States but world-wide. Sites linked to the NNDC are listed at http://www.nndc.bnl.gov/usndp.

The use of electronic data sources is not without its problems by virtue of the enormity of the resource. To be supplied with an excess of data can cause confusion to those who are not sufficiently expert in their use. When applying nuclear techniques to practical problems, the importance of a sound understanding of the scientific principles cannot be over stated. Scientific understanding is the foundation of a knowledge structure while data are the building blocks. A balance must be struck between the two. Nevertheless, electronic data centres represent an almost limitless store of information, making it advisable to refer readers to the World Wide Web, specially so for the latest published data, though the latest data are not necessarily the most useful data.

It is of course necessary to exercise care to ensure that electronically transmitted information is of adequate quality. Clearly National Laboratories and major universities are a first-class source of information. An example is the US National Institute of Science and Technology (NIST) (http://www.physics.nist.gov/phys/refdata/contents). This Institute provides information on topics including:

Physical Reference Data
Constants, Units and Uncertainty
Articles and Publications
Matters of General Interest.

Recognising the rapidly growing importance of Internet sources of information, a listing of sites directly relevant to topics discussed in the book is presented in Table A3.1 below. Information in this table could also help readers to identify references quoted with web or Internet addresses. However, readers should not overlook the extensive and well presented information stored in the references listed at the end of this book. Printed material may not be as easily accessed as the World Wide Web, but it is nearly always as or more informative.

Table A3.1. *Table of Internet sources.*

Topic	Address	Comment
Periodic Table	http://ie.lbl.gov/education/isotopes.htm	The table is linked to a detailed tabulation of the properties of isotopes.
Chart of Nuclides[1]	http://t2.lanl.gov/data/map.html	A comprehensive Chart of the Nuclides. Figure 1.3 is a small selection.
Table of Nuclides	http://www.dne.bnl.gov/CoN/index.html http://ie.lbl.gov/education/isotopes.htm	The table facilitates the search of individual nuclides. Useful information on the production and the applications of the isotope are presented.
Attenuation coefficients	http://physics.nist.gov/PhysRefData/XrayMassCoef/cover.html	Mass attenuation coefficients for elements and a range of compounds of dosimetric interest are presented in tabular and graphical form.
Electron stopping power and range (ESTAR)	http://physics.nist.gov/PhysRefData/Star/Text/ESTAR.html	The stopping power and ranges of electrons in elements and compounds are presented.
X ray data	http://www.csrri.iit.edu/periodic-table.html http://nucleardata.nuclear.lu.se/nucleardata/toi/xraySearch.asp	A useful table linked to the periodic table. It contains X ray emission lines and absorption edges. The table contains an appropriate level of detail for applications. The second entry presents a more detailed listing of X ray and Auger electron levels and intensities.
Decay modes	http://ie.lbl.gov/education/isotopes.htm http://nucleardata.nuclear.lu.se/nucleardata/toi/perchart.html	

Table A3.1. (cont.)

Topic	Address	Comment
Calibration, Units and Uncertainty	http://ts.nist.gov/ts/htdocs/230/233/calibration/uncert/index.html	The US National Institute of Science and Technology has posted a useful discussion of this topic on the Internet.
Food irradiation	http://www.iaea.org/programmes/rifa/icgfi/introduc.htm	A comprehensive listing of the regulatory status of the irradiation of foodstuffs is maintained by the Intergovernmental Committee on Food Irradiation (IGCFI).

Note:
(1) Please note also the subscription site www.nuclides.net and *Nuclides 2000; An Electronic Chart of the Nuclides on Compact Disk* published by the Institute for Transuranium Elements of the Joint Research Centre of the European Commission. The Institute is located at the Kernforshungszentrum, Karlsruhe, Germany.

Appendix 4

Application of tracer techniques to fluid dynamics

A4.1 Introduction

The purpose of this discussion is to introduce readers to basic principles that underpin the applications of tracer techniques to fluid flow investigations.

The general concept of the tracer experiment is illustrated in Figure 8.1. It involves a three-stage process: (a) injecting the tracer T at A; (b) observing the response of the detector to the tracer at B; and (c) using the data to enhance knowledge of the 'system'. As indicated in Section 8.2.1, the term 'system' may be interpreted in a very broad way. However, this discussion will be restricted to water flowing along a pipeline or through a rapidly stirred tank (Figure 8.6(b)).

Let it be assumed that the tracer T is injected into the bulk flow as a short pulse and traverses the cross section B in the time interval between t_0 and $(t_0 + dt)$. A simple expression will be developed for the average transit time or residence time for water particles between the point of injection of the tracer A, and the sampling cross section at B. This expression will be used as the basis for:

- determining the mean residence time and residence time distribution of the tracer 'particles' between A and B;
- establishing the criterion for complete mixing of the tracer with the bulk flow of the water;
- measuring the flow rate of water at B assuming complete mixing has been achieved.

A4.2 Residence time distributions of tracer particles between the points of injection and sampling

Suppose N_T 'particles' of the tracer t are injected as an instantaneous pulse at A, are diluted and dispersed within the system (Figure 8.1) and monitored at B. No two particles have the same transit or residence time between A and B except by chance. It is therefore necessary to describe the arrival times of the tracer particles at B in terms of a distribution of residence times.

Although this presentation is not restricted to a particular geometry, it is helpful to refer to the stirred tank geometry shown in Figure 8.6(b). Let us assume that $dN(t)$ is the number of tracer particles T passing through the exit B within the time interval t to $t + dt$ following the injection at A. The function $dN(t)/dt$ varies with the transit time t and describes the distribution of the transit times between A and B. Closely related to $dN(t)/dt$ is the probability $E(t)$ $(= [dN(t)/dt]/N_T)$ that a tracer particle T will have a residence time between t and $t + dt$, which is referred to as a residence time of t.

Since no two particles have the same transit time, the response $E(t)$ at B to an instantaneous pulse injection at A varies with time. It forms a probability distribution known as the residence time distribution RTD.

Measurement of the residence time distribution (RTD)

It will now be shown that the detector response at B is a direct measure of the RTD of tracer particles under specified conditions.

The function $dN(t)$ defined above is related to the tracer concentration $C(t)$ by the expression

$$dN(t) = C(t)\, dV = C(t)\, Q\, dt, \tag{A4.1}$$

where dV is the volume of water flowing past B in the time interval between t and $t + dt$ and Q is the flow rate which is assumed to be constant.

The fraction of the tracer particles traversing B between t and $t + dt$ may be expressed either as $E(t)dt$ or $dN(t)/N_T$ where N_T is the total number of particles injected, and the probability distribution $E(t)$ is defined above. Hence

$$E(t)dt = dN(t)/N_T = C(t)\, Q\, dt/\, N_T \tag{A4.2}$$

The residence time distribution (RTD) of the particles is the probability distribution $E(t)$ at B (Figure 8.1) following an instantaneous pulse injection of the tracer t at A. It follows from Eq. (A4.2) that, if the bulk flow rate Q is constant, the residence time distribution

$$\text{RTD} = E(t) = C(t)\, Q/N_T = \text{constant} \times C(t) \tag{A4.3}$$

In summary, the response of a detector located at B to an instantaneous injection at A, is the RTD of the tracer in the tank, provided:

- the tracer is injected as an instantaneous pulse, and
- the flow rate Q through the tank is constant.

If there is complete mixing at B, the RTD will be independent of the location of the monitoring point on the cross section. This condition is readily achieved, for instance, if the detector is located at an outlet pipe to the tank (Figure 8.6(b)).

A4.3 Mean residence time for tracer particles between injection and monitoring

As stated above, the fraction of the number of tracer particles t traversing the transect B with a residence time between t and $t + dt$, is $E(t)dt$. Mathematically, the average, or mean residence time MRT, is given by

$$\text{MRT} = \int_0^\infty tE(t)dt. \tag{A4.4}$$

From Eq. (A4.3)

$$\text{MRT} = (Q/N_T)\ \int_0^\infty tC(t)dt \tag{A4.5}$$

where Q (l/s) is the flow rate (assumed constant) and N_T is the total number of tracer particles injected and $C(t)$ is the tracer concentration at B at time t.

If N_T is not known with precision, it may be calculated from the area under the $C(t)$ versus time curve at B (Figure 8.1) provided that complete mixing has been achieved (see below).

Reference was made in Section 8.3.2 and Figure 8.3 to the point to point method of measuring flow rates in pipelines. This method involves monitoring the rate of transport of a radioactive pulse between the responses of the two detectors. Under most circumstances, the pulse maximum locates its position with sufficient accuracy. However, if the pulse shape is not sufficiently well defined, or if the time interval between the two detectors is very short, the pulse maxima cannot be used to accurately define the time interval. Instead, the difference in the mean residence times calculated from Eq. (A4.5) is used.

A4.4 Response to a complex tracer injection

Equations (A4.2) to (A4.5) are valid only for an instantaneous pulse injection of the tracer at A (Figure 8.1). Clearly it is not always desirable to inject the tracer as an instantaneous pulse. Operationally, it may be necessary to add the tracer at a constant rate, as a square pulse or subject to a more complex function. If the tracer input function at A is $C_0(t)$, and the response function at B is $C(t)$, then the input and the response functions are related through the convolution integral

$$C(t) = \int_0^\infty C_0(u)\, h(t-u)\, \mathrm{d}u, \qquad (A4.6)$$

where $h(t-u)$ is the response at B following an instantaneous injection of tracer at A (IAEA, 1990a, Ch. 4). Solving these functions involves a level of detail beyond the scope of this text. Eqation (A4.6) is important as it relates the tracer response function to its behaviour at the input A. In principle the response function at B can be predicted from the input function at A, once the response to the impulse injection is known. Conversely, if $h(t)$ is known, the input function at A can be reconstructed from the observed behaviour at B. This opens up the possibility of assessing sources of pollution from the downstream monitoring of contaminated waterways.

A4.5 Complete mixing

The concept of complete mixing was introduced in Section 8.3.1. If a tracer is injected continuously at a constant rate into a constant flow, complete mixing is achieved downstream of the point at which the tracer concentration is everywhere uniform. This criterion is intuitively simple. By contrast, the criterion for complete mixing following a pulse injection is less intuitive because the downstream concentration of the tracer is nowhere constant, but varies at every sampling point and at every time interval. The equivalent criterion for complete mixing is discussed in Section 8.3.1.

The concept of complete mixing is fundamental to the application of tracer dilution techniques to flow rate measurements. Flow rate measurements are based on considerations of mass balance as expressed in Eq. (A4.1) for a tracer traversing the cross section B following injection at A . The integral form of the equation is

$$N_T = \int_0^\infty \mathrm{d}N(t) = Q\int_0^\infty C(t)\mathrm{d}t. \qquad (A4.7)$$

Equation (A4.7) would only be valid if the value of the integral of the concentration profile were independent of the location of the sampling point on the measurement cross section at B. This is the condition of complete mixing. It may be shown from Eq. (A4.6) that the two criteria for complete mixing are equivalent, namely (a) that uniform concentration is achieved following injection at a constant rate; and (b) the value of the above mentioned integral is independent of the sampling location on the measurement cross section.

A4.6 Flow rate measurements using tracer dilution techniques

As indicated above, tracer dilution techniques can be used to measure flow rates Q, provided complete mixing has been achieved. The fundamental equation is (A4.7). In the following treatment, the number of tracer particles N_T will be replaced by the tracer activity A (= $N_T \times$ decay constant), which is a more conventional nomenclature.

The application of Eq. (A4.7) to the measurement of flow rates involves the evaluation of the integral $\int_0^\infty C(t)\mathrm{d}t$ following the injection of the tracer of activity A. Two methods are commonly used, constant sampling and the total count method.

Total sample method

A sample is taken at a constant rate q (l/s) for a time ΔT which includes the whole pulse. From mass balance considerations,

$$\int_0^\infty C(t)\mathrm{d}t = c\,DT = a/q \qquad (A4.8)$$

where c (Bq/l) is the average concentration of tracer in the total sample and a (Bq) is the corresponding activity.

From Eqs. (A4.7) and (A4.8), $a/q = A/Q$ (or N_T/Q in the 'old' nomenclature), i.e.

$$Q = q\,A/a. \qquad (A4.9)$$

Total count method

If a detector is placed either external to the pipeline or internally within the flow, the registered countrate is proportional to the radioisotope concentration 'seen' by the detector. The total count N recorded by the detector after correction for background and decay is therefore proportional to the integral $\int_0^\infty C(t)\mathrm{d}t$. If F is the proportionality (or calibration) factor, it follows from Eq. (A4.7) that

$$N = F\int_0^\infty C(t)\mathrm{d}t = FA/Q \qquad (A4.10)$$

where the activity A (Bq) = λN_T is replaces N_T and the decay constant λ (s^{-1}) is incorporated into the calibration factor F. The numerical value of F (counts per second registered by the ratemeter for each becquerel per litre in the water) depends critically on the counting geometry and is determined experimentally, as discussed in Section 9.3.2.

From Eq. (A4.10)

$$Q = AF/N. \qquad (A4.11)$$

One of the important features of Eqs. (A4.9) and (A4.11), is that the expressions for the flow rate are independent of the cross sectional area.

A4.7 Residence time distribution (RTD): stirred flow

From Eq. (A4.3), the RTD of tracer particles between the injection and sampling points is given by

$$\mathrm{RTD} = C(t)\,Q/A \qquad (A4.12)$$

where the number of tracer particles N_T and their concentration $C(t)$ are redefined in terms of the activity A (= λN_T Bq) and the activity concentration (Bq/l) . It is assumed that the tracer is injected as an instantaneous pulse and that the flow rate through the system is constant (Figure 8.1).

This discussion will be restricted to calculating the RTD in a rapidly stirred tank, i.e. one in which an injected pulse of tracer is instantaneously dispersed uniformly through the liquid in the tank (Figure 8.6(b)). Under these conditions, a pulse injection of activity A injected at a time t_0 into a reactor of volume V will rapidly assume a concentration C_0 where

$$C_0 = A/V. \tag{A4.13}$$

Assuming that the flow rate through the tank is Q, then the change dC in the tracer concentration between t and $t + dt$ results from the difference between the inflow of tracer (assumed zero) and the outflow ($C(t)Q\,dt/V$), i.e.

$$dC = C(t) - C(t + dt) = [C(t)Q/V]\,dt = [C(t)/\tau]\,dt \tag{A4.14}$$

where the mean residence time $\tau = V/Q$. From Eq. (A4.14), by integration

$$C_1(\theta) = C_0 \exp(-\theta) \tag{A4.15}$$

where $\theta = t/\tau$, and the subscript 1 indicates the concentration profile at the outlet of tank 1.

Hence, the tracer concentration in the first tank will decrease exponentially with time as the labelled material is flushed into the second tank by fresh flow from the upstream inlet port (Figure 8.7). As a consequence, the tracer concentration in the second tank, which is initially zero, will increase through a maximum as tracer flows through the tank. Mathematically, the expression for the tracer concentration at the exit of the second tank is obtained by substituting Eq. (A4.15) (the tracer concentration entering the second tank) into Eq. (A4.14) to obtain

$$C_2(\theta)/C_0 = 4\,\theta\exp(-2\theta) \tag{A4.16}$$

The argument can be repeated for the third and subsequent tanks. In the case of the n th tank the concentration profile may be generalised as

$$C_n(\theta)/C_0 = n^n\,\theta^{n-1}\,e^{-n\theta}/(n-1)! \tag{A4.17}$$

where C_0 is A/V (Eq. (A4.13)), and $(n-1)!$ is the factorial $(n-1)(n-2)(n-3)\dots \times 3 \times 2 \times 1$.

The expressions for $n = 1, 2, 3$ and 5 are depicted in Figure 8.7 as the tracer concentrations at the outlets of the sequence of well mixed adsorption tanks in series. These concentration profiles are to be compared with the behaviour of the extraction (leaching) tanks, which exhibit non-classical behaviour.

Interested readers are referred to Charlton (1986, Ch. 9), and the IAEA Guidebook (IAEA, 1990a) for a detailed derivation of this expression.

Appendix 5

Dispersion processes

A5.1 Introduction

The purpose of this appendix is to provide the interested reader with some of the fundamental concepts underpinning the dispersion of solutes in the aqueous environment. Most practical applications rely on the observations that (a) contaminant transport may be described by advective and dispersive processes which are effectively independent of one another; and (b) following a point source release, concentration profiles often approximate Gaussian distributions (Section 9.3.3). These two observations are a consequence of simple properties of the motion of fluid particles and of the fluctuations in the concentrations of the contaminants associated with the particles.

At any point within the bulk flow, the instantaneous velocity of a fluid particle, u, and the instantaneous concentration of a contaminant, c, may be written as

$$u = \langle u \rangle + u' \tag{A5.1}$$

$$c = \langle c \rangle + c' \tag{A5.2}$$

where $\langle u \rangle$ is the mean velocity in, say, the x or longitudinal direction; u' is the deviation from the mean due to the random motions of the turbulent eddies; $\langle c \rangle$ is the mean contaminant concentration in the fluid particles; and c' is the random deviations of contaminant concentrations in fluid particles about the mean.

The instantaneous rate of transport of the contaminant through a cross section at a particular point in time is the product uc. Hence the mass flux ϕ through a unit cross section averaged over a longer time interval is the average value of this product, i.e. $\langle uc \rangle$. It follows from Eqs. (A5.1) and (A5.2)

$$
\begin{aligned}
\phi = \langle uc \rangle &= \langle [\langle u \rangle + u'][\langle c \rangle + c'] \rangle \\
&= \langle [\langle u \rangle \langle c \rangle + u' \langle c \rangle + \langle u \rangle \, c' + u'c'] \rangle \\
&= \langle \langle u \rangle \langle c \rangle \rangle + \langle u' \langle c \rangle \rangle + \langle \langle u \rangle \, c' \rangle + \langle u'c' \rangle.
\end{aligned}
\tag{A5.3}
$$

Since, by definition, the average values of the fluctuations of particle velocities $\langle u' \rangle$ and contaminant concentrations $\langle c' \rangle$ around the mean are zero,

$$\phi = \langle u \rangle \langle c \rangle + \langle u'c' \rangle, \tag{A5.4}$$

i.e. the flux is the sum of an advection term $\langle u \rangle \langle c \rangle$ and a diffusion term $\langle u'c' \rangle$. This is a justification for the assertion made in Section 9.3.3 that in many investigations, advection and diffusion may be considered separately.

A5.2 Advection and diffusion

Advection is the transport of the solute or contaminant with the bulk flow of the water. Diffusion is the transport of solutes by molecular and turbulent processes which do not lead to net flow. Diffusion processes may be classified as follows.

- *Molecular diffusion* results from the random motion of the solute molecules. In accordance with Fick's law, the net transport of the solute in, say, the longitudinal or x direction $\phi_{\text{mol diff},x}$ is proportional to the concentration gradient $\partial C/\partial x$

$$\phi_{\text{mol diff},x} = -D_{\text{mol diff},x}\, \partial C/\partial x \tag{A5.5}$$

where $D_{\text{mol diff},x}$ is the molecular diffusion coefficient.
- *Eddy diffusion* is a property of turbulent flow, and results from the random motion of small eddies in a solute concentration gradient $\partial C/\partial x$. The expression for the resulting flux $\phi_{\text{eddy diff},x}$ is similar to that for molecular diffusion, namely

$$\phi_{\text{eddy diff},x} = -D_{\text{eddy diff},L}\, \partial C/\partial x \tag{A5.6}$$

where $D_{\text{eddy diff},L}$ is the eddy diffusion coefficient in the longitudinal (x) direction. The eddy diffusion coefficient is typically about 1000 times greater than the molecular diffusion coefficient.
- *Effective diffusion* is the further effect on the distribution of contaminants of large-scale eddies which persist over time scales considerably longer than those characteristic of eddy diffusion. It will not be considered further. Finally, *dispersion* is a generic term that is commonly used to describe turbulent diffusion processes.

A5.3 Gaussian concentration profiles

From considerations of mass balance and the fact that the solute flux is proportional to the concentration gradient (Eq. (A5.6)), it may be shown that

$$\partial C/\partial t + u_x \partial C/\partial x = D_L\, \partial^2 C/\partial x^2. \tag{A5.7}$$

Assuming the instantaneous release of a contaminant from a point source, it may be shown that the solution of Eq. (A5.7) is Gaussian in form.

In the longitudinal, or x direction

$$C(x,t) = C_x^* \exp\left[-(x-u_x t)^2/4D_L t\right] \tag{A5.8}$$

where x is the location on the longitudinal axis; $C(x,t)$ is the concentration of the tracer at the location x and time t; u_x is the velocity of the water in the longitudinal direction; D_L is the longitudinal dispersion coefficient; and C_L^* is the tracer concentration at the centroid of the pulse.

Following the pulse injection of a tracer of activity A_0

$$C_x^* = A_0/2(\pi D_L t)^{1/2} \quad \text{or} \quad A_0/2^{1/2}\pi^{1/2}\sigma_L \tag{A5.9}$$

where the variance (σ_L^2) of the Gaussian distribution is $2D_L t$ (Eqs (9.4) and (6.10)).

Equation (A5.8) may be expressed in three dimensions, namely

$$C(x,y,z,t) = C^*_{x,y,z} \exp\left\{-\left[(x-x_m)^2/4D_L t + (y-y_m)^2/4D_T t\right.\right.$$
$$\left.\left. + (z-z_m)^2/4D_V t\right]\right\} \tag{A5.10}$$

where x,y,z are the locations on the longitudinal, transverse, vertical axes respectively; $C(x,y,z,t)$ is the concentration of the tracer at the location x,y,z and at time t; D_L, D_T, D_V are the longitudinal, transverse or vertical dispersion coefficients; and x_m ($= u_x t$), y_m, z_m is the location and $C^*_{x,y,z}$ the tracer concentration at the centroid of the plume.

Following a pulse injection of tracer (activity A_0)

$$C^*_{x,y,z} = A_0 /[4\pi^{3/2} D_L^{1/2} D_T^{1/2} D_v^{1/2} t^{3/2}] \quad \text{or} \quad A_0 /[2^{1/2}\pi^{3/2} \sigma_L \sigma_T \sigma_V] \quad \text{(A5.11)}$$

where σ_L, σ_T and σ_V are the variances (Eq. (A5.9)) in the longitudinal, transverse and vertical directions respectively.

In some practical applications, the contaminant may be released continuously from, say, a pipeline. A tracer study may involve a longer, or quasi-steady state injection and may be designed to measure the dispersion coefficients in the transverse T (or y) direction and vertical V (or z) direction at different longitudinal distances x from the outlet. In the former case, Eq. (A5.10) may be expressed in the form (Section 9.3.3, Eq. (9.4))

$$C(y,t) = C^*_y \exp\left[-(y-y_m)^2/4D_T t\right] \quad \text{(A5.12)}$$

where y is the location on the transverse axis from the point of maximum concentration y_m; $C(y,t)$ is the concentration of the tracer at the location y and time t; D_T is the transverse dispersion coefficient; and C^*_y is the tracer activity at the maximum concentration (O, Figure 9.3(a)) on the transverse axis.

A similar expression may be written for the vertical dispersion coefficient D_V.

References

Alexiev D. and Butcher K.S.A. (1992) High purity liquid phase epitaxial gallium arsenide nuclear radiation detectors. *Nucl. Instrum. Methods in Physics Research*, **A317**, 111–115.

Alexiev D., Butcher K.S.A. and Williams A.A. (1994) Gamma-ray detectors from CdTe. *J. Crystal Growth*, **142**, 303–309.

Alfassi Z.B. (Ed.) (1990) *Activation Analysis*, Vols. 1 & 2, CRC Press Inc., Boca Raton, Florida, USA.

Bergen Van Der E.A., Jonkers G. and Goethals D. (1989) Industrial applications of positron emission computed tomography. *Nucl. Geophysics*, **3**, 407.

Berger M.J. (1999) *Stopping Power and Range Tables for Electrons, Protons and Helium Ions.* http://physics.nist.gov/PhysRefData/Star/Text/ESTAR.html

Berger M.J. and Seltzer S.M. (1964) *Tables of Energy Losses and Ranges of Electrons and Positrons.* NASA-SP3012, Washington DC, USA.

Bevington P.R. and Robinson D.K. (1992) *Data Reduction and Error Analysis for the Physical Sciences.* McGraw-Hill, New York, USA.

BIPM (1985) *The International System of Units (SI)* [in French and English]. Bureau International des Poids et Mesures, Sevres, France.

Bradley R. (Ed.) (1984) Application Heat-Shrink, Ch. 10, p. 241 and 'Application Sterilisation', Ch. 12, p. 277 in *Radiation Technology Handbook*, Marcel Dekker, New York, USA.

Brissaud I., De Chateau-Thieny A., Frontier J.P. and Lagarde G. (1986). *Application to vulcanic rocks. Analysis of geological standards with PIXE and PIGME techniques, J. Radioanal. and Nuclear Chem.* **102**, 131.

Bull J. (1981) History of Computed Tomography, Ch. 108, pp. 3835–3849 in *Radiology of the Skull and Brain* (Eds. Thomas H. Newton MD and D. Gordon Potts MD), Vol. 5 Technical Aspects of Computed Technology. C.V. Mosby Company, St Louis, Missouri, USA.

Bush, H.D. (1962) *Atomic and Nuclear Physics*, Prentice-Hall Inc., Englewood Cliffs, New Jersey, USA.

Campion P.J. (1959) The standardization of radioisotopes by the β–γ coincidence method. *Int. J. Appl. Radiat. Isotopes*, **4**, 232.

Cember H. (1996) *Introduction to Health Physics*, 3rd edn. McGraw-Hill, New York, USA.

Charlton J.S. (Ed.) (1986) *Radioisotope Techniques for Problem Solving in Industrial Process Plants.* Leonard Hill, Glasgow, UK.

Chu S.Y.F., Ekstrom L.P. and Firestone R.B. (1999) *Table of Radioactive Isotopes.* The Lundt/LBNL Nuclear Data Search, Version 2.0, February 1999, http:// nucleardata.nuclear.lu.se/nucleardata/tol/index.asp

CoN (1981) *Charts of Nuclides.* Kernforschungszentrum Karlsruhe, Germany, Seelmann-Eggebert W., Pfennig G., Munzel H., Klewe-Nebenius H., Institut fur Radiochemie. (5th edition). A later edition (No. 6) was published in 1995 and an amended version of No. 6 in 1998.

Cierjacks E. (Ed.) (1983) *Neutron Sources for Basic Physics and Applications.* Pergamon Press, Oxford, UK.

Clark I.D. and Fritz P. 1997 *Environmental Isotopes in Hydrogeology.* Lewis Publishers, Boca Raton, Florida, USA.

Clarke R.H. (1991) The Causes and Consequences of Human Exposure to Ionising Radiations. *Radiation Protection Dosimetry*, **36**, 73–77.

Cormack A.M. (1963) Representation of a function by its line integral with some radiological applications. *J. Appl. Physics*, **34**, 2722–2727.

Cormack J., Towson J.E.C. and Flower M.A. (1998) Radiation Protection and Dosimetry in Clinical Practice, Ch. 121 in *Nuclear Medicine. Clinical Diagnosis and Treatment*, 2nd edn, (Eds. I.P.C. Murray and R.J. Ell, Churchill Living-stone, Edinburgh, UK.

Cross W.G., Ing H. and Freedman N. (1983) A short atlas of beta-ray spectra. *Phys. Med. Biol.*, **28**(11), 1251–1260.

Crouthamel C.E. (1960) *Applied Gamma-Ray Spectrometry.* Pergamon Press, Oxford, UK. (See also 2nd edn, 1981.)

Cutmore N.G., Evans T. and McEwan A. (1991) On-conveyor determination of moisture in coal. *J. Microwave Power and Electromagnetic Energy*, **26**(4), 237–242.

Cutmore N.G., Howarth W.J., Sowerby B.D. and Watt J.S. (1993) On-line analysis for the mineral industry. *Proc. Australian Institute of Mining and Metallurgy Centenary Conference*, Adelaide, South Australia, pp. 189–197.

Diehl J.F. (1990) *Safety of Irradiated Foods*, Marcel Dekker Inc., New York, USA.

Debertin K. and Helmer R.G. (1988) *Gamma and X-Ray Spectrometry with Semiconductor Detectors*, Elsevier Science Publishers B.V., Amsterdam, The Netherlands.

Duerden P., Cohen D.D., Clayton E., Bird J.R., Ambrose W.R. and Leach B.F. (1979) Elemental analysis of thick obsidian samples by PIXE Spectrometry. *Anal. Chem.*, **51**, 2350.

Easey J. F. (1988) Understanding a Blast Furnace, in *ANSTO Nuclear News*, No. 28 (Ed. G. Carrard), Australian Nuclear Science and Technology Organisation, Menai, NSW, Australia.

Fardy J.J. (1990) Radiochemical Separations in Activation Analysis, Ch. 5, pp. 61–96 in *Activation Analysis*, (Ed. Z. B. Alfassi) Vol. 1, CRC Press, Boca Raton, Florida, USA.

Fisher H.B., List E.J., Koh R.C.Y., Imberger J. and Brooks N.H. (1979) *Mixing in Inland and Coastal Waters* Academic Press Inc., Orlando, Florida, USA.

Firestone, R.B. (1987) *Table of Isotopes*, 8th edn,(Ed. V.S. Shirley), John Wiley and Sons Inc., New York, USA.

Firestone R.B. and Shirley V.S. (1996). *Table of Isotopes*, 8th edn, John Wiley and Sons Inc., New York, USA.

Fricke H. and Hart E.J. (1966) Chemical Dosimetry, Ch. 12 in *Radiation Dosimetry* (Eds. F.H. Attix and W.C. Roesch), Academic Press, New York, USA.

Geiger H. and Werner A. (1924) Die Zahl der vom Radium ausgesandten Alpha Teilchen. *Z. Phys.*, **21**, 187.

Genka T., Kobayashi K., and Hagiwara S. (1987) A calorimeter for the measurement of the activity of tritium and other pure beta emitters. *Appl. Radiat. Isot.*, **38** (10), 845–850.

Genka T., Iwamoto S., Takeuchi N., Uritani A. and Mori C. (1994) Radioactivity measurements of ^{192}Ir wire sources with a microcalorimeter. *Nucl. Instrum. and Methods in Physics Research*, **A339**, 398–401.

Geyh M.A. and Sleicher H. (1990) *Absolute Age Determinations, Physical and Chemical Methods and their Applications*, Springer Verlag, Berlin, Heidelberg.

Gilmore G. and Hemmingway J. (1995) *Practical Gamma-Ray Spectrometry*, John Wiley and Sons, Chichester, England.

Graham J., Higson D.J., Jun J.S., Kobayashi S. and Mitchel R.E.J. (1999) Low Doses of Ionising Radiations at Low Dose Rates. *Radiation Protection in Australasia*, **16**(1), 32–47.

Groenewald W.A. and Wasserman H.J. (1990) Constants for calculating ambient and directional dose equivalents from radionuclide point sources. *Health Physics*, **58**(5), 655–658.

Heath R.L. (1964) *Gamma-Ray Spectrum Catalogue*, 2nd edn, USAEC Report IDO-16880-1, National Technical Information Service, Springfield, VA, USA.

Heath R.L. (1974) *Gamma-Ray Spectrum Catalogue*, 3rd edn, Report ANCR-1000-2, National Technical Information Service, Springfield, VA, USA.

Heron M.M. (1987) Future applications of elemental concentrations from geophysical logging. *Nucl. Geophysics*, **1**, 197–211.

Hey T. and Walters P. (1987) *The Quantum Universe*, Cambridge University Press, Cambridge, UK.

Hills A.E. (1999) *Practical Guidebook for Radioisotope Based Technology in Industry*, Published as IAEA Report IAEA/RCA, RAS/8/078, IAEA, Vienna, Austria.

Hino Y., Kawada Y. and Nazaroh (1996) Improved leakage current compensation for pressurised ionisation chambers. *Nucl. Instrum. Methods in Physics Research*, **A369**, 391–396.

Hore-Lacy I. (Ed.) (1999), *Nuclear Electricity*, 5th edn, Uranium Information Centre (Australia) Ltd, Melbourne 3001, Victoria, Australia.

Hubbell J.H. (1982) Photon mass attenuation and energy-absorption coefficients from 1 keV to 20 MeV. *Int. J. Appl. Radiat. Isotopes*, **33**, 1269.

Hughes J. (2000) 1932: the *annus mirabilis* of nuclear physics? *Physics World*, July, pp. 43–48.

Hunter E., Maloney P., Bendayan M. and Silver M. (1993) *Practical Electron Microscopy: A Beginner's Illustrated Guide*. Cambridge University Press, Cambridge, UK.

IAEA (1970) *Neutron Fluence Measurements*, Technical Report Series No. 107, International Atomic Energy Agency, POB 100, A-1400, Vienna, Austria.

IAEA (1971) *Nuclear Well Logging in Hydrology*, Technical Report Series No. 126, Article by J.A. Czubek,Ch. II B, pp. 26–31, International Atomic Energy Agency, POB 100, A-1400, Vienna, Austria.

IAEA (1973) *Neutron Well Logging in Hydrology*, Technical Report Series No. 126, Ch. II C, pp. 32–46, International Atomic Energy Agency, POB 100, A-1400, Vienna, Austria.

IAEA (1983) *Guidebook on Nuclear Techniques in Hydrology*, with contributions by J. Guizerix and T. Florkowski. Stream Flow Measurements Ch. 5, pp. 65–91, and G.S. Tazioli and A. Caillot, pp. 93–101, International Atomic Energy Agency, POB 100, A-1400, Vienna, Austria.

IAEA (1986) *Radiation Protection Glossary*. IAEA Safety Series No. 76, International Atomic Energy Agency, POB 100, A-1400, Vienna, Austria.

IAEA (1987) *Handbook on Nuclear Activation Data*, Technical Report Series No. 273, International Atomic Energy Agency, POB 100, A-1400, Vienna, Austria.

IAEA (1990a) *Guidebook on Radioisotope Tracers in Industry*, Technical Report Series No. 316, International Atomic Energy Agency, POB 100, A-1400, Vienna, Austria.

IAEA (1990b) *Prompt Gamma Neutron Activation Analysis in Borehole Logging and Industrial Process Control*, IAEA-TECDOC-537, with contributions from J. Charbucinski, J.A. Aylmer, P.L. Eisler and M. Borsaru 'Quantitative and Qualitative Applications of the Neutron–Gamma Borehole Logging' pp. 85–103; and from J.L. Mikesell, F.E. Senftle, R.N. Anderson and M. Greenberg 'Elemental Concentrations in Igneous Rocks Determined by High-Resolution Gamma-Ray Spectrometry and Applications to Petrochemical Problems' pp. 105–134, International Atomic Energy Agency, POB 100, A-1400, Vienna, Austria.

IAEA (1993) *Handbook on Nuclear Data for Borehole Logging and Mineral Analysis*, IAEA Technical Report Series No. 357, International Atomic Energy Agency, POB 100, A-1400, Vienna, Austria.

IAEA (1995) *Use of Nuclear techniques in Studying Soil Erosion and Erosion*, contributions by P. E. Aun and J.V. Bandeira 'The Role of Nuclear Techniques in Sedimentological Studies and some Applications in Latin America' pp. 29–97; and J.C. Ritchie and C.A. Ritchie '^{137}Cs use in Erosion and Sediment deposition Studies: Promises and Problems', pp. 111–124.

IAEA (1996a) *Basic Safety Standards for Protection against Ionising Radiation and for the Safety of Radiation Sources*, IAEA Safety Series No. 115, International Atomic Energy Agency, POB 100, A-1400, Vienna, Austria.

IAEA (1996b) *Regulations for the Safe Transport of Radioactive Materials* Safety Standard Series ST-1 International Atomic Energy Agency, POB 100, A-1400, Vienna, Austria.

IAEA (1997) *The Thin Layer Activation Method and its Application to Industry*, IAEA TECDOC – 924, International Atomic Energy Agency, POB 100, A-1400, Vienna, Austria.

ICRP (1983) *Radionuclide Transformations, Energies and Intensities of Emissions* ($\sim$ 1200 decay schemes). International Commission on Radiological Protection, Publication 38, Volumes 11–13, Elsevier Science Ltd, Kidlington, Oxford, UK.

ICRP (1991) *Publication 60*. 1990 Recommendations of the International Commission on Radiological Protection, Volume 21, pp. 1–3, Elsevier Science Ltd, Kidlington, Oxford, UK.

ICRP (1997) *Radiation Protection, Publication 77*, Radiological Protection Policy for the Disposal of Radioactive Waste. Elsevier Science Ltd, Kidlington, Oxford, UK

ICRU (1993) *Quantities and Units in Radiation Protection Dosimetry*, ICRU Report 51, International Commission on Radiation Units and Measurements, Bethesda, Maryland, USA (see also ICRU, Report 33, 1980).

Jaklevic J.M., Loo B.W. and Goulding F.S. (1977) Photon Induced X ray Fluorescence Analysis using Energy Dispersive Detector and Dichotomous Sampler Ch. 1, pp. 3–18 in *X ray Fluorescence Analysis of Environmental Samples* (Ed. T.G. Dzubay), Ann Arbor Science, Michigan, USA.

James W.D. (1990) Activation Analysis of Coal and Coal Effluents, Ch. 8, pp. 359–376 in *Activation Analysis* (Ed. Z.B. Alfassi) Vol. 2, CRC Press Inc., Boca Raton, Florida, USA.

Jonkers G., Van Der Bergen E.A. and Vermont P.A. (1990) Industrial applications of a gamma ray camera system. *Applied Radiations and Isotopes*, **41** (10/11), 1023–1031.

Kawada Y., Mogi J. and Hino Y. (1996) Reduction of the γ-ray background when using 32 keV X rays from ^{137}Cs. *Nucl. Instrum. Methods in Physics Research*, **A369**, 671–675.

Keay C. (1999) Effects of Ionising Radiation. *The Physicist*, **39** (5), 192–195.

Kirkup L. (1994) *An Introduction to the Analysis and Presentation of Data*, John Wiley and Sons, Milton, Queensland 4064, Australia.

Knoll G.F. (1989) *Radiation Detection and Measurement*, 2nd edn, John Wiley and Sons, New York, USA.

Kruger P. (1971) *Principles of Activation Analysis*, Wiley-Interscience, New York, USA.

Lachance G.R. and Claisse F. (1995) *Quantitative X ray Fluorescence Analysis: Theory and Applications*, John Wiley and Sons, New York, USA.

Lagoutine F., Coursol N. and Legrand J. (1975 to 1987, periodically updated and continued) *Table de Radionucleides*, C.E.A, B.N.M., Centre d'Etudes Nucleaires de Saclay, F91190, Gif-sur-Yvette. See also LMRI 1988, *Radioactivity Standards*, Catalogue of the Laboratoire de Metrologie des Rayonnements Ionisants, Saclay, France.

L'Annunziata M.F. (Ed.) (1998). *A Handbook of Radioactiovity Analysis*, Academic Press, New York, USA.

Lederer C.M. and Shirley V.S. (Eds.) (1978) *Table of Isotopes*, 7th edn, John Wiley and Sons, New York, USA.

Le Guennec B., Alquier M., Santos-Cottin H. and Margrita R. (1978) Analysis of a solid liquid pipe flow by a cross correlation method. *International Journal of Multiphase Flow*, **4**, 511.

Lewis R. (1997) *Dispersion in Estuaries and Coastal Waters*, John Wiley and Sons, Chichester, UK.

Longworth G. (Ed.) (1998) *The Radiochemical Manual*, AEA Technology plc, Analytical Services Group, Harwell, Oxfordshire, UK.

Luckey T.D. (1998) Impressions of the IAEA/WHO Conference on Low Doses from Ionising Radiations, Nov. 1997, Seville, Spain. *Radiation Protection Management*, **15**, 1, Guest Editorial.

Mann W.B., Ayres R.L. and Garfinkel S.B. (1980) *Radioactivity and Its Measurement*, 2nd edn, Pergamon Press, Oxford, UK.

Mann W.B., Rytz A. and Spernol A. (1991) *Radioactivity Measurements, Principles and Practice*, Elsevier Science, Oxford, OX5 1GB, UK.

Martin A. and Harbison S.A. (1996) *An Introduction to Radiation Protection*, 4th edn, Chapman and Hall, London, UK.

Martz H.E., Azevedo S.G., Brase J.M., Waltjon, K.E. and Schneberk D.J. (1990) Computed Tomography Systems and Their Industrial Applications, *Appl. Radiations and Isotopes*, **41**, 943–961.

McCulloch D.B. and Wall T. (1976) *A method for measuring neutron absorption cross sections of soil samples for calibration of the neutron moisture meter.* Nucl. Instrum. Methods, **137**, 577–581.

McEwan A.C. (1999) Is Cosmic Radiation Exposure of Air Crews amenable to Control? *Radiation Protection in Australasia*, **16** (3), 21–25.

McLaughlin W.L., Boyd A.W., Chadwick K.H., McDonald J.C. and Miller A. (1989) *Dosimetry for Radiation Processing*, Taylor and Francis, London, UK.

Miller W., Alexander R., Chapman N., McKinley I. and Smellie J. (1994) Natural Analogue Studies in the Geological Disposal of Radioactive Waste, *Elsevier Studies in Environmental Science 57*, Amsterdam, The Netherlands.

Morgan I.L. (1990) Real time digital gauging for process control. *Appl. Radiation and Isotopes*, **41**, 935–942.

Morris N.D. (1992) *Personal Radiation Monitoring and Assessment of Doses Received by Radiation Workers* ARL/TR107, Australian Radiation Laboratory, Yallambie, Victoria, Australia.

Mozumber A. (1999) *Fundamentals of Radiation Chemistry*, Academic Press, New York, USA.

NCRP (1985) *A Handbook of Radioactivity Measurements Procedures*, Report No. 58, NCRP Publications, Bethesda, MD 20814, USA.

Neiler J.H. and Bell P.B. (1966) The Scintillation Method, Ch. 5 in *Alpha-, Beta- and Gamma-Ray Spectroscopy* (Ed. K. Siegbahn) Vol. 1, North-Holland Publishing Amsterdam, The Netherlands.

Parratt L.G. (1961) *Probability and Experimental Errors in Science*, John Wiley and Sons, New York, USA.

Parsons P.A. (1992) Evolutionary adaptation and stress, the fitness gradient. *Evolutionary Biology*, **26**, 191–223.

Parsons P.A. (1999) Low level exposure to ionising radiation: Do ecological and evolutionary considerations imply phantom risks? *Perspectives in Biology and Medicine*, **43** (1), 57–68.

Peisach M. (1990) Prompt activation analysis with charged particles, Ch. 3, pp. 143–218 in *Activation Analysis*, (Ed. Z.B. Alfassi) CRC Press Inc., Boca Raton, Florida, USA.

Peng C.T. (1977) Sample preparation in liquid scintillation counting, RCC Review 17, The Radiochemical Centre, Amersham, UK.

Pennington W. (1972) Recent Sediments of Windermere. *Freshwater Biology*, **3**, 363–382.

Pennington W., Cambray R.S., and Fisher E.M., (1973) Observations of Lake sediments using fallout ^{137}Cs as a tracer. *Nature*, **242**, 324–326.

Playford K., Lewis G.N.L., Carpenter R.C. (1992) *Radioactive Fallout in air and rain: Results to the end of 1990*, Report. AEA-EE-0362 27, AEA, Harwell, Oxfordshire, UK.

Pritchard T.R., Lee R.S. and Davison A. (1993) Sydney Deepwater Outfalls: In situ Observations of Plume Characteristics *Proceedings Eleventh Conference on Coastal and Ocean Engineering*, Townsville, pp. 53–58.

Romans L.E. (1995) *Introduction to Computed Tomography*, Williams & Wilkins, Baltimore, Maryland, USA.

Rytz A. (1983) The international reference system for activity measurements of γ-emitting nuclides. *Int. J. Appl. Radiat. and Isotopes*, **34**, 1047.

Santry D.C., Bowes G.C. and Munzenmayer K. (1987) Precision electronics for ionisation chamber measurements. *Int. J. Appl. Radiat. and Isotopes*, **38**, 879.

Seltzer S.M. and Berger M.J. (1982) Energy losses and ranges of electrons and positrons. *Int. J. Appl. Radiat. Isotopes*, **33**, 1189.

Shani G. (1990) Activation analysis with isotopic sources, Ch. 5, pp. 239–297 in *Activation Analysis* (Ed. Z.B. Alfassi) Vol. 2, CR Press Inc., Boca Raton, Florida, USA.

Shapiro J. (1972) *Radiation Protection, A Guide for Scientists and Physicians*. Harvard University Press, Cambridge, Mass., USA (see also 2nd edn, 1981).

Shleien B. (Ed.) (1992) *The Health Physics and Radiological Health Handbook*. Revised Edition, Scinta Inc, Silver Spring, Maryland, USA. See also Shleien B., Slaback L.A. Jr. and Birky B. (Eds.), (1997) 3rd edn, William and Wilkins, POB 23291, Baltimore, Maryland, USA.

Simmons J.A. and Watt D.E. (1999) *Radiation Protection: A Radical Reappraisal*, Medical Physics Publishing, Wisconsin, USA.

Sowerby B.D. (1985) Determination of ash, moisture and specific energy in coal. IAEA Advisory Group Meeting on gamma, x-ray and neutron techniques in the Coal Industry, Vienna 4–7 Dec. 1984, pp. 131–147.

Spiegel S. (1999) *Outline of Statistics*, 3rd edition, McGraw Hill, Sydney, Australia.

Standards Australia (1998) *Safety in Laboratories, Ionizing Radiations*, Australian Standard AS2243.4–1998, Standards Association of Australia, Homebush, NSW, Australia.

Sumner T.J., Grant S.M, Alexiev D. and Butcher K.S.A. (1994) LPE GaAs as an X ray detector for astronomy. *Nucl. Instrum. Methods in Physics Research*, **A348**, 518–521.

Tabata Y., Ito Y. and Tangawa S. (Eds.) (1991) *CRC Handbook of Radiation Chemistry*, CRC Press, Boca Raton, Florida, USA.

Tuniz C., Bird J.R., Fink D. and Herzog G.F. (1998) *Accelerator Mass Spectrometry, Ultrasensitive Analysis for Global Science*, CRC Press, Boca Raton, Florida, USA.

Turner J.E. (1986) *Atoms, Radiations and Radiation Protection*, Pergamon Press, New York, USA.

UIC (1998) Information Leaflets (regularly updated), Uranium Information Centre (Australia) Ltd, GPO Box 1649N, Melbourne 3001, Victoria, Australia.

Umezawa H., Kaneko Y. and Shimizu M. (1996) Supply and waste management of radioisotopes in Japan. *J. Radioanal. Nucl. Chem.*, **205**(1), 21–33.

UN (1956) Proceedings of the International Conference on the Peaceful Uses of Atomic Energy, Geneva 1955, **15**, *Applications of Radioactive Isotopes and Fission Products in Research and Industry*. Contributions by J.W. Watkins and H.N. Dunning 'Radioactive Isotopes in Petroleum Production Research' pp. 32–37 ; and by D.E. Hull and B.A. Fries (1956) 'Radioisotopes in Petroleum Refining, Research and Analysis' pp. 199–210, United Nations, New York, USA.

UN (1958) Proceedings of the International Conference on the Peaceful Uses of Atomic Energy, Geneva 1958, **19**, *The Use of isotopes: Industrial Use*, United Nations, New York, USA.

UNSCEAR (1988) *Sources, Effects and Risks of Ionising Radiations*, United Nations Scientific Committee on the Effects of Atomic Radiations, New York, USA.

Wapstra A.H. and Audi G. (1985) The 1983 Atomic Mass Evaluation. *Nucl. Physics*, **A432**, 1.

Wasserman H.J. and Groenewald W.A. (1988) Air Kerma Rate Constants for Radionuclides. *Europ. J. Nucl. Med.*, **14**, 569–571.

Watt D.E. and Ramsden D. (1964) *High Sensitivity Counting Techniques*, Pergamon Press, Oxford, UK.

Watt I.M. (1997) *The Principles and Practice of Electron Microscopy*, Cambridge University Press, Cambridge, UK.

Watt J.S. (1972) Radioisotope detector-radiator assemblies in X ray fluorescence analysis for copper and zinc in iron-rich minerals. *Int. J. Applied Rad. Isotopes*, **23**(4), 257–264.

Watt J.S. (1973) Radioisotope on-stream analysis. *Atomic Energy in Australia*, **16**(4), 3–19.

Wellington S.L. and Vinegar H.J. (1987) X ray computerised tomography. *J. PET Technology*, **39**, 885.

Wells P., Davis J.R., Seunderman B., Shadbolt P.A., Benci N., Grant J.A., Davies D.R. and Morgan M.J. (1992) A simple transmission X ray micro-tomography Instrument. *Nucl. Instrum. Methods in Physics Research*, **B72**, 261–270.

Williams D.B. and Carter C.B. (1996) *Transmission Electron Microscopy: A Textbook for Materials Science*, Plenum Publications, New York, USA.

Wilson B.J. (Ed.) (1966) *The Radiochemical Manual*, 2nd edn. The Radiochemical Centre, Amersham, UK (see Longworth G. (Ed.) (1998) for revised and updated edition).

Wilson R.D., Stronswold D.C., Mills W.R. and Cook T.K. (1989) Porosity logging using epithermal neutron lifetime Monte-Carlo simulation. *Nucl. Geophysics*, **3**, 323–334.

Wischmeier W.H. (1976) Use and Misuse of the Universal Soil Loss Equation *J. Soil Water Conservation*, **31**, 5.

Wischmeier W.H. and Smith D.D. (1965) Predicting Rainfall Erosion Losses for for Cropland east of the rocky Mountains. *Agr. Handbook No. 262*, US Department of Agriculture, Washington DC, USA.

Woods R.J. and Pikaev A.K. (1993) *Applied Radiation Chemistry: Radiation Processing*, John Wiley and Sons, New York, USA.

Zatz L.M. (1981) Basic Principles of Computed Tomography, Ch. 109 in *Radiology of the Skull and Brain* (Eds. T.H. Newton and D.G. Potts), Vol. 5, Technical Aspects of Computed Tomography', C.V.Mosby Company, St Louis, Missouri, USA.

Index

331